Wider den Reduktionismus

Oliver Passon · Christoph Benzmüller
(Hrsg.)

Wider den Reduktionismus

Ausgewählte Beiträge zum Kurt
Gödel Preis 2019

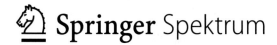 Springer Spektrum

Hrsg.
Oliver Passon
Fakultät für Mathematik und
Naturwissenschaften
University of Wuppertal
Wuppertal, Nordrhein-Westfalen
Deutschland

Christoph Benzmüller🆔
Department of Maths and CS
Freie Universität Berlin
Berlin, Deutschland

ISBN 978-3-662-63186-7   ISBN 978-3-662-63187-4   (eBook)
https://doi.org/10.1007/978-3-662-63187-4

Die Deutsche Nationalbibliothek verzeichnet diese Publikation in der Deutschen Nationalbibliografie; detaillierte bibliografische Daten sind im Internet über http://dnb.d-nb.de abrufbar.

Planung/Lektorat: Lisa Edelhäuser
Springer Spektrum ist ein Imprint der eingetragenen Gesellschaft Springer-Verlag GmbH, DE und ist ein Teil von Springer Nature.
Die Anschrift der Gesellschaft ist: Heidelberger Platz 3, 14197 Berlin, Germany

Vorwort

Im Jahr 2019 vergab der *Kurt Gödel Freundeskreis* (Berlin)[1] in Kooperation mit der Bergischen Universität Wuppertal zum ersten Mal den mit insgesamt 15.000 EUR dotierten *Kurt Gödel Preis*.[2] In der Ankündigung formulieren René Talbot und Hans Schwarzlow für den *Freundeskreis:*

> Dieser Preis wird im Rahmen eines Essaywettbewerbs für die beste Frage vergeben, die Reduktionisten beantworten müssten, dies jedoch nicht können und warum.*

Und in der Fußnote präzisieren sie weiter:

> „Antireduktionistisches Wissen ist die philosophische Antwort auf die Frage, warum die Welt nicht reduktionisch zu erklären ist. Und zwar mit einer so basalen Argumentation, dass jeder Versuch zum Scheitern verurteilt ist, der mit einem Verweis auf mögliche, später noch zu erbringende Beweise reduktionistische Erklärungsversuche zu retten sucht."

Mit diesen provokanten Formulierungen forderte der Kurt Gödel Freundeskreis also bewusst zu einer pointierten und streitlustigen Auseinandersetzung auf. Die formalen Bedingungen für die Teilnahme waren gering: Der Beitrag sollte mindestens 4000 Zeichen umfassen und in deutscher oder englischer Sprache verfasst sein.

Aus den zahlreichen (anonymisierten) Einsendungen wählte eine unabhängige Fachjury bestehend aus Eva-Maria Engelen (Konstanz), Christoph Benzmüller (Berlin) und Oliver Passon (Wuppertal) zunächst eine Shortlist preiswürdiger Essays aus. Aus diesen einigte man sich auf:

1. Preis für Jesse M. Mulder („The limits of reductionism: thought, life, and reality")

---

[1]Siehe https://kurtgoedel.de. Dort finden sich auch Anmerkungen dazu, warum der bedeutende Logiker Kurt Gödel (1906–1978) ein passender Namensgeber für diesen Preis ist; siehe dazu aber auch den Beitrag von Reinhard Kahle (Kap. 10).

[2]Die Idee zum *Kurt Gödel Preis* war bereits 2018 entstanden. Das Vorhaben wurde dann im Februar 2019 im Rahmen des an der FU Berlin ausgerichteten Workshops „Kurt Gödel: Philosophical Views" weiter konkretisiert (cf. http://www.christoph-benzmueller.de/2019-Goedel).

2. Preis für George F. R. Ellis („Why reductionism does not work")
3. Preis für Tim Lethen („Monads, Types, and Branching Time – Kurt Gödel's approach towards a theory of the soul")

Der vorliegende Band enthält neben diesen ausgezeichneten Beiträgen weitere Essays aus der Shortlist. Die ausgewählten Artikel diskutieren die Preisfrage aus unterschiedlichen Perspektiven – vor allem allgemein philosophische, naturwissenschaftlich-physikalische oder stärker logisch-mathematische Argumente wurden gewählt.[3]

Dabei ist es kaum notwendig hervorzuheben, dass die Auswahl eines Beitrages – trotz der oben zitierten Formulierung in der Ausschreibung – keine Einschätzung zur Unwiderlegbarkeit der darin aufgestellten Behauptungen und Aussagen umfassen kann. Dafür ist zum einen der Begriff „Reduktionismus" zu schillernd – und kann sich dadurch einer Widerlegung entziehen.[4] Andererseits ist es gerade Merkmal von philosophischen (bzw. wissenschaftlichen) Positionen, dass sie zwar Geltung beanspruchen, aber eben keine Letztgültigkeit (vgl. dazu den Beitrag von Hueske, die genau dieses Argument aber auch *gegen* den Reduktionismus wendet).

Aber wie lauten nun die Fragen, deren befriedigende Beantwortung dem Reduktionismus nicht gelingen soll, und gegen welche Lesart des Reduktionismus wird argumentiert?

Wenden wir uns hier zunächst den philosophischen Argumenten zu. **Martin Breul** (Kap. 1) bemerkt in seinem Aufsatz zunächst, dass „klassische" Kandidaten für Irreduzibilität wie die Intentionalität mentaler Zustände oder Qualia in weit verzweigte Debatten führen. Im Gegensatz dazu schlägt er vor, mit der Kausalität einen grundlegenderen Begriff in das Zentrum des Diskurses zu rücken. Nach Breul zeichnen sich reduktionistische Forschungsprogramme dadurch aus, dass sie Kausalbeziehungen mit „basalem ontologischen Status" aufweisen. Er untersucht die Probleme eines solchen Verständnisses von Kausalität und schlägt die konkrete Frage „Was hat den Großen Brand von London verursacht?" vor.

Hanna Hueske (Kap. 2) unterscheidet zwischen einem „materialistischen" und einem „logischen" Reduktionismus. Während der erstere alle Eigenschaften komplexer Systeme auf ihre physischen Bestandteile zurückführen möchte, versucht der andere, alle Phänomene durch Explikation ihrer logischen Struktur vollständig zu erfassen. Hueske argumentiert in ihrem Aufsatz diskurstheoretisch und formuliert für beide Varianten des Reduktionismus folgendes Problem: „Wie kann der Reduktionismus sich selbst als Position rechtfertigen?"

---

[3]Nach diesen nicht immer ganz trennscharfen Kategorien ist auch das Buch in drei Teile gegliedert. Innerhalb dieser ist eine alphabetische Reihenfolge der Autorinnen und Autoren gewählt worden.

[4]Die Mehrdeutigkeit der Begriffe „Reduktion" und „Reduktionismus" drückt sich auch darin aus, dass alle Autorinnen und Autoren dieses Bandes *unterschiedliche* Argumente vortragen, um jeweils mehr oder weniger spezifische Varianten des „Reduktionismus" zu widerlegen.

Tim Lethen (Kap. 3) formuliert als einziger Autor keine explizite Frage. Dieser Beitrag stellt frühe Überlegungen Gödels zur Monadologie und dem metaphysischen Programm von Leibniz dar, und er verwendet dazu bisher unveröffentlichte Quellen aus Gödels Nachlass.

Jesse Mulder (Kap. 4) erwähnt eine Reihe von möglichen „lokalen" und „globalen" Reduktionismen. Sogar ein Idealismus nach Berkeley könne als solcher bezeichnet werden, denn schließlich reduziere dieser alle materiellen Objekte auf bloße Wahrnehmungsinhalte. Mulder macht jedoch den (weit verbreiteten) Physikalismus zum Gegenstand seiner Kritik, dem zufolge „alles über physikalische Tatsachen superveniert". Sein Aufsatz führt in einem Crescendo durch die Fragen „What is thought?", „What is life?" und schließlich „What is reality?"

Michał Pawłowski (Kap. 5) siedelt seine Diskussion in der Philosophie des Geistes an. Hier betrifft der Reduktionismus (den auch Pawlowski mit dem Physikalismus identifiziert) die Reduzierbarkeit mentaler Zustände auf ihre neuronalen Korrelate. Pawlowski argumentiert (Hueske nicht unähnlich), dass rationale Begründung und Wahrheitsanspruch für den Reduktionisten in einem Spannungsverhältnis stehen. Seine Frage lautet somit: „How can you believe in the truth and rational justification of your view at the same time?"

Unter den stärker naturwissenschaftlich orientierten Beiträgen diskutiert **George Ellis** (Kap. 6) ebenfalls die Kausalitätsproblematik. Ein weiterer Schlüsselbegriff seines Aufsatzes ist jedoch die „Emergenz". Dieses Konzept behauptet grob gesprochen das mögliche Auftreten *neuartiger* Eigenschaften eines Systems aus dem *Zusammenwirken* seiner Elemente. Emergente Eigenschaften sind dabei gerade nicht auf die Eigenschaften der Bestandteile reduzierbar. Ellis umfangreicher Beitrag diskutiert eine Fülle von Beispielen für dieses Phänomen[5] und sucht damit die folgende These zu begründen: „Reductionists cannot answer why strong emergence (unitary, branching, and logical) is possible, and in particular why abstract entities such as thoughts and social agreements can have causal powers."

Rico Gutschmidt (Kap. 7) wendet sich in seinem Essay der Frage zu, ob innerhalb der Physik höherstufige Theorien tatsächlich eliminativ auf grundlegendere reduziert werden können. Im Verhältnis zwischen Newtonscher und Quantenmechanik sieht er hier fundamentale Probleme. Seine Frage lautet somit: „Wie lässt sich die makroskopische Welt in Begriffen der Quantenmechanik beschreiben, ohne dabei Newtonsche Konzepte zu verwenden?"

Eine dritte Klasse von Beiträgen wendet sich dem Gegenstand eher logisch-mathematisch zu. **Jean-Yves Béziau** (Kap. 8) rekonstruiert den Reduktionismus als die Forderung, für einen Gegenstandsbereich Prinzipien anzugeben, aus denen seine vollständige und logische Herleitung gelingt. Er untersucht in seinem Beitrag die Grenzen einer solchen Axiomatisierbarkeit, und die Frage an den

---

[5]Dabei behandelt er auch so alltägliche Vorgänge wie den Schluckauf (Abschn. 6.3.6) und das Stricken (Abschn. 6.4.2).

Reduktionismus formuliert Béziau in gewisser Weise schon im Titel seines Beitrages: „Is there an axiom for everything?"

Marco Hausmann (Kap. 9) fasst den Reduktionismus als These über
bestimmte (vollständige) Erklärungsansprüche auf. Mit den Mitteln der Modallogik führt er diese Annahme zu einem Widerspruch. Daraus folgt bei ihm die
Existenz (wenigstens) einer „unerklärten Wahrheit". Die zugehörige Frage
(„Kannst du mir diese Wahrheit erklären?") ist für den Reduktionisten also unbe-
antwortbar.[6]

Reinhard Kahle (Kap. 10) schließlich erweist in seinem Beitrag dem Namens-
patron des Preises Reverenz. Er zeigt, wie schon aus dem ersten Unvollständigkeitssatz ein Argument gegen den „mathematischen Reduktionismus" folgt. Die
Frage etwa, wie reelle Zahlen auf natürliche Zahlen zurückgeführt werden können,
lässt sich mit Mitteln der finiten Mathematik *nicht* beantworten.

Der besondere Charme dieser Aufsatzsammlung liegt aber nicht nur in den verschiedenen Perspektiven, sich dem Gegenstand anzunähern. Ebenfalls versammelt
dieser Band Autorinnen und Autoren, die an ganz verschiedenen Stellen ihrer
Laufbahn stehen – von Studierenden im Bachelor- oder Promotionsstudium über
Postdocs bis zu arrivierten Hochschullehrenden.

Zusammen mit der Internationalität der Teilnehmenden sowie der Zwei-
sprachigkeit dieser Veröffentlichung ist so ein wirklich facettenreiches Werk
entstanden, bei dessen Lektüre wir den Leserinnen und Lesern viel Vergnügen
wünschen!

Wuppertal und Berlin   Oliver Passon
Januar 2021   Christoph Benzmüller

---

[6]Bei diesem Existenzbeweis bleibt natürlich offen, wie diese Frage konkret lautet.

Inhaltsverzeichnis

Herausgeber- und Autorenverzeichnis

Über die Herausgeber

Christoph Benzmüller ist Professor für KI/Informatik, Logik und Mathematik an der Freien Universität Berlin. Er war der erste UNA-Europa-Gastprofessor, und er kooperiert derzeit mit einem Berliner Start-up-Unternehmen. Aktuelle Forschungsarbeiten adressieren normatives Schließen und rationales Argumentieren im bzw. mit dem Computer, universelle Wissensrepräsentation, computationale Metaphysik und formale Grundlagen und Mechanisierung der Mathematik. Zu früheren Stationen seiner Karriere gehören die Universitäten in Luxemburg, Stanford (USA), Cambridge (UK), Birmingham (UK), Edinburgh (UK) sowie die Carnegie Mellon University (USA) und die Universität des Saarlandes.

Oliver Passon ist Privatdozent an der Bergischen Universität Wuppertal und lehrt Physik und ihre Didaktik. Seine Interessens- und Arbeitsgebiete umfassen Goethes Farbenlehre, die phänomenologische Optik sowie die Didaktik, Geschichte und Philosophie der Quantentheorie. Zu seinen jüngsten Buchveröffentlichungen zählen (zusammen mit C. Friebe, M. Kuhlmann, H. Lyre, P. Näger und M. Stöckler) *The Philosophy of Quantum Physics* (Springer, 2018) und (zusammen mit T. Zügge und J. Grebe-Ellis) *Kohärenz im Unterricht der Elementarteilchenphysik* (Springer, 2020).

Autorenverzeichnis

Jean-Yves Béziau is a professor and researcher of the Brazilian Research Council (CNPq) at the University of Brazil in Rio de Janeiro. He works in the field of logic – in particular, paraconsistent logic and universal logic. He holds a PhD in Philosophy from the University of Sao Paulo and a PhD in Logic and Foundations of Computer Science from Denis Diderot University (Paris). Béziau is the editor-in-chief of the journal Logica Universalis and of the South American Journal of Logic as well as of the Springer book series Studies in Universal Logic. Currently he is president of the Brazilian Academy of Philosophy.

Martin Breul ist Systematischer Theologe und Religionsphilosoph. Nach dem Studium der Philosophie, Katholische Theologie und Anglistik in Köln, Münster und Belfast promovierte er 2015 zum Dr. phil. mit einer Arbeit über das Verhältnis von Religion und Politik. 2018 folgte die Promotion zum Dr. theol. mit einer Arbeit über Jürgen Habermas' Diskurstheorie. Derzeit arbeitet er am Institut für Katholische Theologie der Universität zu Köln im DFG-Projekt „Die theologische Relevanz von Michael Tomasellos evolutionärer Anthropologie". Seine Forschungsschwerpunkte liegen im Bereich Religion und Politik, theologische und evolutionäre Anthropologie, Gotteslehre.

George F. R. Ellis, FRS, is Emeritus Distinguished Professor of Complex Systems in the Department of Mathematics and Applied Mathematics at the University of Cape Town. He is one of the world's leading researchers in general relativity theory and cosmology and co-authored *The Large Scale Structure of Space-Time* (1973) with Cambridge physicist Stephen Hawking.

Born in Johannesburg, Ellis graduated from the University of Cape Town in South Africa and received his PhD at Cambridge University. His research interests cover amongst other things the history and philosophy of cosmology, complex systems and the relation of science to religion.

George Ellis was a vocal opponent of apartheid during the National Party reign in the 1970s and 1980s and awarded the Order of the Star of South Africa by Nelson Mandela, in 1999.

Rico Gutschmidt studierte Mathematik, Physik und Philosophie und wurde 2009 mit einer Arbeit zum Reduktionsproblem in der Physik an der Universität Bonn im Fach Philosophie promoviert. Im Anschluss hat er an der TU Dresden zu Martin Heidegger gearbeitet und dessen späte Philosophie im Kontext von Denkfiguren negativer Theologie neu interpretiert. Nach der Habilitation in Dresden im Jahre 2015 arbeitete er in Chicago, Hamburg und Valparaiso. Derzeit ist er akademischer Mitarbeiter an der Universität Konstanz; seine jüngeren Forschungsinteressen umfassen Wittgenstein, Skeptizismus und transformative Erfahrungen in der Philosophie.

Marco Hausmann ist Promotionsstipendiat der Studienstiftung des deutschen Volkes und hat einen Lehrauftrag an der Ludwig-Maximilians-Universität München. Er arbeitet an einer Promotion im Fach Philosophie zum Thema Freiheit und Determinismus. Seine Forschungsgebiete umfassen Metaphysik, Logik und Sprachphilosophie. Zu seinen jüngsten Veröffentlichungen zählen „The Consequence of the Consequence Argument" (*Kriterion,* 2019), „Against Kripke's Solution to the Problem of Negative Existentials" (*Analysis,* 2019), „The Consequence Argument Ungrounded" (*Synthese,* 2018).

Hanna Hueske studiert seit 2018 Philosophie und Mathematik an der Leibniz Universität Hannover. Ihre Interessenschwerpunkte sind die Philosophie der Antike, insbesondere Platon, sowie die Sprachphilosophie und Philosophie der Mathematik des 20. Jahrhunderts. Der Beitrag in diesem Band stellt ihre erste Veröffentlichung dar.

Reinhard Kahle ist Carl Friedrich von Weizsäcker-Stiftungsprofessor für Theorie und Geschichte der Wissenschaften an der Universität Tübingen. Zuvor war er Professor für Mathematik in Coimbra und an der Universidade Nova de Lisboa. Er ist Mitglied der Académie Internationale de Philosophie des Sciences. Zu seinen Forschungsinteressen gehören Beweistheorie und die Geschichte und Philosophie der modernen mathematischen Logik, insbesondere im Umfeld der Hilbertschule. Er hat zusammen mit Kollegen mehr als zehn Bücher und Sondernummern herausgegeben, unter anderem *Gentzen's Centenary: The quest for consistency* (Springer, 2015) und *The Legacy of Kurt Schütte* (Springer, 2020), beide zusammen mit Michael Rathjen.

Tim Lethen, equipped with a degree in mathematics and computer science, has studied the Gabelsberger shorthand system in order to be able to read the personal and scientific notes in Kurt Gödel's Nachlass, kept at the Institute for Advanced Study in Princeton. Since 2018, Tim has been working for the Godeliana project based in Helsinki, Finland, led by Jan von Plato. As part of this project, he has transcribed many of Gödel's notebooks, including several books on theology and on the foundations of quantum mechanics. Tim is mainly interested in Gödel's formal approach to metaphysics and theology, on which he has written several papers.

Jesse M. Mulder obtained his PhD in philosophy from Utrecht University in 2014. His dissertation sketches a broadly Aristotelian, anti-reductionist version of conceptual realism. Subsequently, he worked as a postdoctoral researcher at Utrecht's computer science department, in the context of Prof. Dr. Jan Broersen's ERC-funded project „Responsible Intelligent Systems". Since 2016, he is assistant professor at Utrecht's Department of Philosophy and Religious Studies. His philosophical interests continue to include metaphysics, philosophy of logic, philosophy of biology, philosophy of mind, and action theory. How these various fields hang together is the topic of his current individual research project, funded by the Dutch national funding agency NWO, entitled Unifying Metaphysical Pluralism (December 2017–August 2021).

Michał Pawłowski, born in 1996 in Warsaw, is currently a graduate student at the University of Warsaw, last year also on an exchange at the University of Konstanz. Michal holds a MA degree in philosophy (June 2020) as well as a BA in economic sciences and philosophy. His recent philosophical interests include self-referential argumentations and limits of knowledge in metaphysics, early Ludwig Wittgenstein and philosophy of culture. Although his academic background involves most notably the analytic tradition, he really enjoys linking it with the tools offered by the 'continental' paradigm in philosophy.

Teil I
Philosophische Perspektiven

Kausalität als antireduktionistisches Hausmittel oder: Was hat den Großen Brand von London verursacht?

Martin Breul

Im September 1666 brach in London ein Feuer aus, das als ‚Großer Brand von London‘ in die Geschichte eingehen sollte. Das Feuer wütete zehn Tage lang und zerstörte 80 % der Stadt – 400 Straßen, 13.200 Häuser und 87 Kirchen fielen den Flammen zum Opfer. Nach dem Brand wurde fieberhaft nach einem Schuldigen gesucht. Obwohl diverse Verschwörungstheorien kursierten – eine beliebte Verschwörungstheorie bestand darin, den katholischen Jesuitenorden der Brandstiftung zu bezichtigen –, konnte am Ende gezeigt werden, dass das Feuer in einer Bäckerei an der Pudding Lane ausgebrochen war, da ein Bäcker am Abend zuvor Reste von glimmender Glut im Ofen übersehen hatte, die über Nacht zunächst die Backstube in Brand setzten. Von dieser brennenden Backstube aus entfaltete das Feuer seine zerstörerische Kraft.

Philosophisch lässt sich aus dieser Tragödie einiges lernen. In meinen Augen ist eine vielleicht etwas überraschende Lehre, dass sich in der Frage nach der Ursache nach dem Großen Brand von London ein anti-reduktionistisches Argument versteckt. Die philosophischen Debatten zwischen Reduktionisten und Anti-Reduktionisten drehen sich in den letzten Jahren – wie schon in den Jahrhunderten zuvor – um recht ‚klassische‘ Kandidaten für irreduzible Phänomene.[1] Anstelle in die weit verzweigten Diskurse um die (Ir-)Reduzibilität von Qualia, Intentionalität, Moral etc. einzusteigen, ist es in meinen Augen sinnvoll, die Kategorie der Kausalität in den Blick zu nehmen, da sich bereits an dieser zunächst unverdächtigen Stelle die Unmöglichkeit eines Reduktionismus zeigt. Kurzum, die Frage, die Reduktionisten beantworten müssten, dies jedoch nicht können, lautet: Was hat den Großen Brand von London verursacht?

---

[1] Eine ausführliche Zusammenstellung dieser irreduziblen Phänomene findet sich exemplarisch in Nagel (2013).

M. Breul (✉)
Institut für Katholische Theologie, Universität zu Köln, Köln, Deutschland

© Der/die Autor(en), exklusiv lizenziert durch Springer-Verlag GmbH, DE, ein Teil von Springer Nature 2021
O. Passon und C. Benzmüller (Hrsg.), *Wider den Reduktionismus,*
https://doi.org/10.1007/978-3-662-63187-4_1

Für reduktionistische Forschungsprogramme ist die Annahme von Kausalbeziehungen mit basalem ontologischen Status ein zentrales Element. Bestimmte gesetzmäßige Zusammenhänge müssen reduzierbar sein auf deterministische Ursache-Wirkungs-Beziehungen. Der damit implizierte metaphysische Begriff von Kausalität setzt voraus, dass eine bestimmte Ursache hinreichend ist, um eine bestimmte Wirkung eintreten zu lassen. Der amerikanische Philosoph Hilary Putnam nennt eine solche hinreichende Ursache auch „totale Ursache" (Putnam 1993, S. 182).

Im Normalfall werden kausalitätstheoretische Erwägungen genutzt, um Reduktionismen zu begründen – es wäre kontraintuitiv zu glauben, dass ein Ereignis auf eine Vielzahl von Ursachen zurückgeht. Putnam versucht nun, den Spieß umzudrehen und zu zeigen, dass genau dieser Reduktionismus in Fragen von Ursache-Wirkungs-Beziehungen problematisch ist. Sein Kernargument besagt, dass es sich bei ‚totalen Ursachen' um Fiktionen des menschlichen Geistes handelt, denn: In Ursache-Wirkungs-Beziehungen werden stets bestimmte Ursachen ausgewählt, um ein Ereignis zu erklären. Im strengen Sinne kann es hinreichende Ursachen nicht geben, da das Angeben von Ursachen oder Ursachenbündeln immer schon eingelassen ist in das jeweilige explikative Interesse, das mit der Frage nach einer spezifischen Ursache verbunden ist.[2]

Daher bleibt die Suche nach allen ursächlich wirksamen Faktoren für ein Ereignis notwendig unvollständig. Um diese ‚totale Ursache' zu finden, müsste sich der jeweilige Forscher über alle Kontexte, in denen er forscht, erheben und das Ganze der Welt in den Blick nehmen können. Jeder Forscher ist aber notwendig Teil des Ganzen, sodass das Ganze nicht mehr als Ganzes in den Blick kommen kann. Das reduktionistische Forschungsprogramm einer Rückführung aller Wissenschaften auf eine Superwissenschaft, die es schafft, die fundamentalen Strukturen des Seins zu erhellen, scheitert schon an Überlegungen zur kausalen Struktur der Welt.

Dies kann man für unseren Kontext schön am Beispiel des Großen Brandes von London verdeutlichen: Ein ursächlich wirksames Ereignis für die Feuersbrunst besteht darin, dass der Bäcker an der Pudding Street an einem Samstagabend vergaß, sicherzustellen, dass sich keine glimmende Glut mehr im Ofen befand. Im Normal-

---

[2]Diesen Zusammenhang zwischen der Möglichkeit von Erkenntnis und dem Einfluss von Forschungsinteressen einerseits hat auch Jürgen Habermas schon in seiner Auseinandersetzung mit dem Positivismus in den frühen 1960er-Jahren hervorgehoben. Es ist sowohl falsch, das Wissen der Naturwissenschaften mit Erkenntnis überhaupt zu identifizieren, da es eine Vielzahl weiterer, nichtnaturwissenschaftlicher oder nichtempirischer Kontexte gebe, in denen Erkenntnisse gewonnen werden, als auch methodisch prekär, die alle Einzelwissenschaften leitenden Erkenntnisinteressen nicht in die notwendige Selbstreflexion jeder Wissenschaft einzubeziehen (vgl. Habermas 1968). Sowohl für Putnam als auch für Habermas steht dabei eine pragmatistische Philosophie Pate, da sie eine Vielzahl ihrer Intuitionen und Argumente aus den Schriften von Charles Sanders Peirce und John Dewey ableiten. Dieser rote Faden des philosophischen Pragmatismus zieht sich durch das gesamte Werk Habermas' und findet sich auch an einer zentralen Stelle seines letzten großen Werks *Auch eine Geschichte der Philosophie*: „[D]as pragmatistische Argumente [behauptet], dass es sinnlos ist, für die Objektivität der Erkenntnis einen ortlosen Standpunkt oder Blick von nirgendwo zu postulieren. Vielmehr gehört es zum Begriff der Erkenntnis, dass sie ‚unsere' Erkenntnis ist" (Habermas 2019, S. 747).

fall würden wir dieses Ereignis auch als (juristisch relevante) Ursache des Feuers betrachten, jedoch ist es nicht die ‚totale Ursache' des Großen Brandes von London. Ursächlich wirksam waren auch die Tatsachen, dass die große Mehrheit der Häuser Londons im Jahre 1666 aus Holzfachwerk erbaut waren; dass zu dieser Zeit üblicherweise jedes weitere Stockwerk leicht überhängend auf das darunterliegende gebaut wurde, was das Übergreifen des Feuers auf weitere Häuser ermöglichte; der vorangehende heiße und trockene Sommer sowie der starke und zugleich warme Südwestwind in den Tagen der Feuersbrunst etc. Wenn man sich darüber hinaus einmal vorstellt, dass nichtmenschliche Lebensformen, die überhaupt nicht bekannt mit den Hintergrundbedingungen menschlicher Existenz sind – beispielsweise Außerirdische –, eines Tages auf die Welt kommen und vom Großen Brand von London hören, könnten auch ganz andere Ursachen in den Blick kommen. Die Außerirdischen könnten beispielsweise das ihnen völlig unbekannte Vorhandensein einer ausreichenden Menge Sauerstoff in der Erdatmosphäre als Ursache für den Großen Brand von London benennen. Daher lässt sich mit Putnam festhalten: „[W]as für den einen die Hintergrundbedingung darstellt, [kann] sehr leicht für einen anderen die ‚Ursache' sein. Was eine ‚Ursache' oder ‚Erklärung' ist oder nicht ist, *hängt vom Hintergrundwissen und unserem Grund ab, diese Frage zu stellen*." (Putnam 1993, S. 184, Hervorhebung M. B.).

Ein weiteres Beispiel, das Putnam ebenfalls zur Illustration dieser kontextuellen Gebundenheit der wissenschaftlichen Untersuchung von Ursache-Wirkungs-Beziehungen gibt, besteht im Fall, dass im Studierendenwohnheim eine Person A um Mitternacht splitterfasernackt im Zimmer einer anderen Person B gefunden wird. Eine mögliche ‚totale Ursache' dieser Situation könnte man so beschreiben: „Seine [Person A; M.B.] Anwesenheit in dem Raum zum Zeitpunkt ‚Mitternacht − ϵ' (wobei ϵ so klein ist, dass er sich zwischen Mitternacht − ϵ und Mitternacht weder anziehen noch den Raum verlassen kann, ohne sich schneller als Licht zu bewegen) wäre eine totale Ursache dafür, dass er sich zu Mitternacht nackt im Zimmer […] befindet. Niemand würde das aber als die ‚Ursache' bezeichnen." (Putnam 1993, S. 183).[3] Der Grund dafür, dass niemand diese Erklärung ernsthaft als Ursache für die Anwesenheit von Person A im Zimmer von Person B betrachten würde, besteht darin, dass Ursachen stets die Aufgabe haben, relevante Erklärungen von Zuständen zu sein. Relevanz ist aber eine Kategorie, die von Erkenntnisinteressen, Handlungszusammenhängen und Forschungskontexten abhängig ist, sodass das Definieren einer Ursache lebensweltlich gebunden bleibt. Mit anderen Worten: Das Konzept von Ursache-Wirkungs-Beziehungen und damit von Kausalität ist abhängig von unseren explikativen Bedürfnissen, die wiederum in verschiedenen lebensweltlichen Kontexten stark variieren können und die nicht objektiv definierbar sind. Explanatorische Argumente sind immer schon eingelassen in bestimmte epistemische und lebens-

---

[3]In Putnams Beispiel ist Person A ein Professor, Person B ist eine Studentin – dies erklärt das falsche Personalpronomen im Ausdruck ‚seine Anwesenheit'. Zugleich zeigt Putnams unbekümmerte Konstruktion des Beispiels – Person A ist mächtig und männlich, Person B ist Person A untergeordnet und weiblich – eine fehlende Sensibilität für die Dekonstruktion von Machtstrukturen. Daher habe ich es hier vorgezogen, das Beispiel anonymisiert darzustellen.

weltliche Hintergrundbedingungen, und die explikative Kraft einer Erklärung ist abhängig von unseren Erkenntnisinteressen.

Daher lässt sich für eine pragmatistisch fundierte Kritik des Reduktionismus auf Basis einer Analyse der Kategorie der Kausalität festhalten, dass

> jeder explanatorisch nutzbare Kausalitätsbegriff eine epistemische Komponente enthält und auf pragmatisch gerechtfertigte Erklärungszusammenhänge bezogen ist. […] Kausalität ist nicht ein geistunabhängiger metaphysischer Klebstoff zwischen Ereignissen, sondern eine Kategorie, die zutiefst an menschliche Handlungszusammenhänge gebunden ist. (Brüntrup 2018, S. 149 f.)

Diese Deflationierung der Kategorie der Kausalität zeigt in meinen Augen, dass die Welt nicht reduktionistisch zu erklären ist und auch jeder Versuch, im Verweisen auf eine zukünftige ideale Naturwissenschaft den Reduktionismus zu plausibilisieren, zum Scheitern verurteilt ist. Der Grund dafür besteht in der untrennbaren Verwobenheit der menschlichen Erkenntnis mit Handlungskontexten und Erkenntnisinteressen. Das grundlegende Problem des Reduktionismus, welches sich am Beispiel der Kausalität schön illustrieren lässt, besteht in der Annahme, dass dem menschlichen Erkenntnisapparat ein Durchgriff auf die objektive Realität, d. h. die ontologisch basalen Strukturen der Welt, möglich ist. Ein reduktionistisches Verständnis von ‚kausalen Zusammenhängen‘ ist ein Paradebeispiel, um die vermeintliche ‚inhärente Struktur der Wirklichkeit‘, die der Reduktionist postuliert, aufzuzeigen. Mit Hilary Putnam gesprochen lässt sich die Annahme einer objektiven, quasi ‚eingebauten‘ kausalen Struktur in die Realität als die Annahme einer ‚Fertigwelt‘ bezeichnen. Diese Annahme ist für den Reduktionismus wiederum erforderlich, wenn man davon ausgeht, dass Wahrheit in einer unperspektivischen denkerischen Kopie ebendieser Fertigwelt besteht.[4]

---

[4]Diese Einsicht ist im Übrigen nicht auf die Philosophie beschränkt, sondern wurde auch von einigen Theologinnen und Theologen formuliert. So hat Karl Rahner, einer der einflussreichsten katholischen Theologen des 20. Jahrhunderts, schon 1955 in erkenntniskritischer Absicht festgehalten: „Der Mensch ist in seiner Erkenntnis nicht eine photographische Platte, die gleichgültig und ungewandelt einfach registriert, was je gerade im einzelnen, abgetrennten Augenblick auf sie fällt. Er muss vielmehr, schon um bloß zu verstehen, was er sieht oder hört, reagieren, Stellung nehmen, die neue Erkenntnis in einen Zusammenhang bringen." (Rahner 2004, S. 24). Die Unmöglichkeit der ‚Abbildung‘ der Wirklichkeit an sich bzw. des unverstellten Zugangs zur ‚Wahrheit an sich‘ ist dabei nichts, was Rahner als Spezifikum der Theologie betrachtet; vielmehr trifft diese Einsicht auch für ihn explizit auf alle Wissenschaften bzw. menschlichen Erkenntnisbemühungen zu. Rahner nennt exemplarisch einen Physiker, der einen physikalischen Vorgang im isolierenden Experiment betrachtet: Eine adäquate, d. h. erschöpfende Beschreibung des Geschehens ist aufgrund der Interdependenzen der Wirklichkeit unmöglich – würde man versuchen, das Geschehen als Beschreibung der Realität an sich verständlich zu machen, müsste man beanspruchen, dass man „sich selbst mit seiner eigenen physikalischen Wirklichkeit auf einen Punkt stellen könnte, der schlechthin und in jeder Beziehung außerhalb dieses Kosmos läge: ein Ding der Unmöglichkeit." (Rahner 2004, S. 23).

Es ist nun wichtig zu sehen, dass, bei aller vermeintlichen Affinität zur modernen (Natur-) Wissenschaft, ein solcher Reduktionismus bzw. die Annahme einer solchen Fertigwelt eine metaphysische Theorie über die Beschaffenheit des Ganzen der Welt ist, die mit empirischen Methoden gar nicht prüfbar sein kann. Es kommt also auf die philosophischen Argumente an, die für diese Position ins Feld geführt werden können. Die Wahrheit des Reduktionismus lässt sich nicht naturwissenschaftlich oder empirisch prüfen, sondern ist letztendlich eine weltanschauliche ‚Wette'.

Angesichts dieser Überlegungen lässt sich der zentrale Einwand gegen das Eingehen dieser Wette auf die Wahrheit eines solchen metaphysischen Reduktionismus so formulieren: Der Reduktionismus kann seine Versprechen nicht einlösen, weil er *methodisch* unhaltbar ist. Seine Erkenntnisansprüche setzen implizit die Möglichkeit eines epistemischen Gottesstandpunktes voraus, die aber reine Fiktion ist. Daher scheitert der Reduktionismus „an der methodischen Fiktion eines exklusiven ‚Blicks von nirgendwo', der sich einer problematischen Entkoppelung der objektivierenden Perspektive des naturwissenschaftlichen Beobachters von der eines Teilnehmers an der Forschungspraxis verdankt" (Habermas 2009, S. 273). Letztlich ist damit ein jedes reduktionistisches Weltbild – und umfassende Reduktionismen sind keine wissenschaftlichen Forschungsprogramme, sondern Weltbilder – eine metaphysische Überhöhung wissenschaftlicher Erkenntnisse zu einer allumfassenden Welterklärung. Es gilt stattdessen, sich abzufinden mit der „Pluralität von einigen tief verankerten welterschließenden Perspektiven" (Habermas 2009), weshalb eine reduktive, d. h. singuläre und allumfassende Beschreibung der Wirklichkeit eine unerreichbare Fiktion darstellt.

Abschließend lässt sich also festhalten, dass man für die Widerlegung eines umfassenden Reduktionismus nicht notwendig in die Untiefen philosophischer Debatten über die (Nicht-) Reduzierbarkeit von Qualia, Intentionalität, Moral etc. hinabsteigen muss. Es ist hinreichend zu zeigen, dass der Reduktionismus die unhintergehbare Perspektivität menschlicher Erkenntnis verkennt und damit in seinen Erkenntnisansprüchen über das hinausschießt, was sich sinnvollerweise überhaupt sagen lässt. Der Reduktionismus beruht im Letzten auf der Fiktion eines erkenntnistheoretischen Gottesstandpunktes, von dem aus das Ganze des Seins erforschbar oder auch nur neutral beobachtbar ist. Damit aber wird eine bestimmte (meist physikalische) Beschreibung der Wirklichkeit einseitig verabsolutiert und verkannt, dass jede Forschende immer Teil des Ganzen ist und ihr Blick auf das Ganze notwendig ausschnitthaft bleiben muss. Dies lässt sich beispielhaft an der Kategorie der Kausalität zeigen, die nicht eine eigenständige Größe in der objektiven Welt, sondern eine von der Forscherin selbst abhängige Deutung bestimmter Ereignisse ist. Daher ist eine zentrale Frage, die Reduktionisten beantworten müssten, es aber nicht können, die Frage nach der objektiven hinreichenden Ursache eines Ereignisses. Auf unser Beispiel angepasst: Eine Frage, auf die es nicht die eine wahre Antwort gibt, d. h. die von Reduktionisten nicht beantwortet werden kann, ist die auf den ersten Blick völlig banale Frage: „Was hat den Großen Brand von London verursacht?".

Literatur

Brüntrup, G. (2018). *Philosophie des Geistes. Eine Einführung in das Leib-Seele- Problem*. Stuttgart: Kohlhammer.

Habermas, J. (1968). *Erkenntnis und Interesse*. Frankfurt a. M.: Suhrkamp.

Habermas, J. (2009). „Das Sprachspiel verantwortlicher Urheberschaft und das Problem der Willensfreiheit: Wie läßt sich der epistemische Dualismus mit einem ontologischen Monismus versöhnen?" In J. Habermas (Hrsg.), *Philosophische Texte: Band 5. Kritik der Vernunft* (S. 271–341). Frankfurt a. M.: Suhrkamp.

Habermas, J. (2019). *Auch eine Geschichte der Philosophie, Bd. 2: Vernünftige Freiheit. Spuren des Diskurses über Glauben und Wissen*. Berlin: Suhrkamp.

Nagel, T. (2013). *Geist und Kosmos. Warum die materialistische neodarwinistische Konzeption der Natur so gut wie sicher falsch ist*. Berlin: Suhrkamp.

Putnam, H. (1993). „Warum es keine Fertigwelt gibt". In von V. C. Müller (Hrsg.), *Von einem realistischen Standpunkt: Schriften zu Sprache und Wirklichkeit* (S. 174–202). Reinbek bei Hamburg: Rowohlt. (Original: Why there isn't a ready-made world. *Synthese, 51*(1982), 141–168).

Rahner, K. (2004). „Die Assumptio-Arbeit von 1951 mit den Änderungen bis 1959". In *Karl Rahner Sämtliche Werke: Band 9. Maria, Mutter des Herrn, Mariologische Studien (Bearbeitet von Regina Pacis Meyer)*. Freiburg im Breisgau: Herder.

Reduktionismus im Diskurs

2

Hanna Hueske

In philosophischen und wissenschaftlichen Debatten werden reduktionistische Positionen seit der Antike vertreten. Obwohl die häufigste reduktionistische Position wohl der Materialismus ist, ist für eine erschöpfende Darstellung der Problematik reduktionistischer Ansätze eine Betrachtung zweier auf den ersten Blick sehr verschiedener Positionen notwendig: Auf der einen Seite wird ein materialistischer Reduktionismus vertreten, der die Eigenschaften komplexer Systeme zurückführen will auf ihre physikalische, chemische oder biologische Zusammensetzung aus Teilchen oder Organen. Auf der anderen Seite steht eine Art von Reduktionismus, den ich logischen Reduktionismus nennen möchte, da er den Anspruch erhebt, Phänomene durch die Explikation ihrer logischen Strukturen vollständig zu erfassen. Inwiefern er dem materialistischen Reduktionismus gegenübersteht und doch zum selben Ergebnis führt, wird sich im Folgenden zeigen.

Sobald eine Person, eine Personengruppe oder eine philosophische Schule eine Form des Reduktionismus vertritt und dies äußert, wird der Reduktionismus zu einer Position. Eine Position im philosophischen Diskurs zeichnet sich durch die gleichzeitige Annahme zweier sich nur auf den ersten Blick widersprechender Haltungen aus:[1] Erstens muss sie einen Geltungsanspruch erheben, also davon ausgehen, dass es eine Wahrheit gibt und sie selbst diese Wahrheit vertritt. Das führt unmittelbar zu einem Rechtfertigungszwang: Die Position muss durch Argumente für andere nachvollzieh-

---

[1]Das folgt aus der Auffassung von Philosophie als „begründende Rede, die durch Explikation impliziter Voraussetzungen danach strebt, begründete Rede zu sein und für Andere und von deren Einstimmung her gilt" (Zorn 2016, S. 45). Warum Philosophie immer auch in dieser Auffassung bezeichnet werden kann, begründet Zorn in seiner Arbeit ausführlich.

H. Hueske (✉)
Leibniz Universität Hannover (Studentin des fächerübergreifenden Bachelor Philosophie),
Hannover, Deutschland
E-mail: hanna.hueske@posteo.de

© Der/die Autor(en), exklusiv lizenziert durch Springer-Verlag GmbH, DE, ein Teil von
Springer Nature 2021
O. Passon und C. Benzmüller (Hrsg.), *Wider den Reduktionismus,*
https://doi.org/10.1007/978-3-662-63187-4_2

bar gemacht werden können, um sich im Diskurs behaupten zu können. Zweitens
muss die Position die Möglichkeit ihres Scheiterns aber immer mitberücksichtigen.
Trotz ihres Geltungsanspruchs muss sie bereit sein, gehaltvolle Gegenargumente als
solche anzuerkennen und ihren Anspruch möglicherweise aufzugeben.[2]

Der materialistische Reduktionismus versucht, jede logische Struktur, also auch
die Rahmenbedingungen einer philosophischen Diskussion, auf materiale Bestandteile zurückzuführen. Er kann daher nicht über die logischen Voraussetzungen der
Rede über die Welt sprechen, sondern nur über die Welt als Gesamtheit materialistischer Einzelbestandteile. Das führt dazu, dass es zunächst kein anderes Kriterium
für die Wahrheit einer Aussage über die Existenz einer Sache gibt, als die physische
Existenz dieser Sache. Da auch die menschliche Vernunft kein Kriterium mehr ist,
muss schließlich sogar jeder Geltungsanspruch aufgegeben werden, denn selbst der
Begriff der Wahrheit ist nicht mehr haltbar – die Unterscheidung von wahr und falsch
ist bereits eine logische Unterscheidung und hat in einer Welt, die nur aus Materie
besteht, keinen Gehalt.

Dieser Reduktionismus kann tatsächlich von keiner anderen Position aus widerlegt werden, weil er selbst die Notwendigkeit von Rechtfertigung gar nicht akzeptiert.
Jede Position, die dem materialistischen Reduktionismus vorwirft, falsch zu sein, verwendet ein Strohmann-Argument, das dem Reduktionismus einen Anspruch unter-
stellt, den er nicht erheben kann. Sobald ein Vertreter des materialistischen Reduktionismus diese Tatsache erkennt, steht er vor einem Dilemma: Entweder vertritt er
seine Position weiterhin und begeht damit einen performativen Selbstwiderspruch,
indem er einen Geltungsanspruch erhebt, der niemals erfüllt werden kann, weil jeder
Geltungsanspruch eine Rechtfertigungspflicht erfüllen muss. Oder er verzichtet auf
den Geltungsanspruch und nimmt damit implizit einen Skeptizismus an, der selbst
die Äußerung seiner Position sinnlos macht.

Der logische Reduktionismus reduziert die Erkenntnis der Wirklichkeit nicht auf
die Erforschung ihrer materialen Bestandteile, sondern betrachtet die Wirklichkeit
ausschließlich aus der Perspektive der menschlichen Vernunft, die sie als absoluten
(also von externen Kriterien losgelösten) Maßstab setzt. Auf den ersten Blick scheint
ein Mensch mit dieser Einstellung bestens geeignet für den philosophischen Diskurs,
der von Rationalität als Maßstab ausgehen muss. Allerdings wählt der Diskurs ja
gerade diesen Maßstab, weil die Vernunft als Logos von allen Diskursteilnehmern
geteilt werden kann. Eine Haltung, die Rationalität nicht als Gemeinsames anerkennt,
führt somit unweigerlich in einen philosophischen Dogmatismus: Sobald jemand, der
den logischen Reduktionismus als Position vertritt, keinen anderen Maßstab annimmt
als das eigene Urteilsvermögen, ist seine Teilnahme am Diskurs sinnlos.[3] Er muss
dann voraussetzen, dass alle anderen seiner Position immer zustimmen müssen oder

---

[2]Vgl. Zorn (2016, S. 248): „Ein philosophischer Logos ist so im Vorhinein nur als Geltungsbe-
hauptung zu sehen und nicht schon als Geltungssetzung."
[3]Beispiele für diese Form des Reduktionismus lassen sich u. a. in Platons Darstellung antiker Sophistik finden (vgl. Platon 2004, S. 40). In seiner Einleitung schreibt Helmut Meinhardt: „Der Weg von
dieser radikalen Zuversicht in die Kraft des Denkens zum radikalen Relativismus der Sophistik ist
kürzer, als man zunächst vermutet. Wenn nur das Bestand hat, wahr ist, was durch das Denken

schon zugestimmt haben, weil sie richtig ist – allerdings ist seine Position für andere nicht verifizierbar, da kein anderer Teilnehmer am Diskurs Zugriff auf die Vernunft des Dogmatikers hat. Damit der Reduktionismus seinen Status als Position behalten kann, muss er also annehmen, dass es einen anderen Maßstab außerhalb der eigenen Vernunft, des eigenen Logos, gibt. Jeder logische Reduktionismus widerspricht damit, sobald er sich äußert, im operativen Rahmen des Diskurses der Position, die er inhaltlich zu vertreten behauptet.

Die wichtigste Frage, die der Reduktionismus beantworten können müsste, lautet also: Wie kann der Reduktionismus sich selbst als Position rechtfertigen? Wir haben gesehen, dass jeder Versuch, die Frage aus reduktionistischer Sicht zu beantworten, an seinen eigenen Voraussetzungen scheitern muss. Sobald der Reduktionismus sich selbst als Position in einem Diskurs behaupten will, hat er bereits operativ Ansprüche erhoben, die er auf inhaltlicher Ebene immer wieder aufheben muss. Materialistischer Reduktionismus als Skeptizismus und logischer Reduktionismus als Dogmatismus teilen die Eigenschaft, dass sie die Debatte, die sie gewinnen wollen, selbst verunmöglichen. Schon in der Philosophie, die sich selbst als rechtfertigende Rede versteht und damit auf Geltungsansprüche angewiesen ist, kann der Reduktionismus sich nicht behaupten und wird damit auch für die Verwendung als Erklärungsgrundlage anderer Wissenschaften, die Anspruch auf vollständige Erklärung der Wirklichkeit erheben, uninteressant. Schließlich wird ersichtlich, warum keine Einzelwissenschaft die Wirklichkeit vollständig erklären kann: Jede Naturwissenschaft würde, sollte sie diesen Anspruch an sich selbst haben, skeptizistisch in einen materialistischen Reduktionismus übergehen; jede Geisteswissenschaft dogmatisch in einen logischen Reduktionismus.

Literatur

Platon. (2004). *Der Sophist (übers. von H Meinhardt)*. Stuttgart: Reclam.
Zorn, D.-P. (2016). *Vom Gebäude zum Gerüst: Entwurf einer Komparatistik reflexiver Figurationen in der Philosophie*. Berlin: Logos.

---

erschlossen wird, dann wiederum ist offensichtlich alles konsistent und wahr, was durch Denken dargelegt wird."

Monads, Types, and Branching Time—Kurt Gödel's Approach Towards a Theory of the Soul

3

Tim Lethen

3.1 Introduction

In 1935, Kurt Gödel wrote two consecutive notebooks entitled *'Physik—Quantenmechanik I'* and *'Physik—Quantenmechanik II,'*[1] which—to this day—remain completely unpublished but have now been entirely transcribed from the Gabelsberger shorthand system by the present author. The books contain Gödel's thoughts, ideas, and questions about the foundations of quantum mechanics, carefully devised into a single list of about 340 items. Only one year later, Gödel wrote a third notebook,[2] entitled *'Aflenz 1936 (Analysis, Physik),'* compiled on the basis of the earlier two books: Whereas some of the items are simply just copied, others are carefully revised and sometimes extended. Also, many of the original notes are completely dropped, with the original order of the items being retained.

In a reply to a questionnaire sent to Gödel by Burke D. Grandjean in 1974, published by Hao Wang (1987), Gödel himself explains that he studied Leibniz between 1943 and 1946, adding that 'the greatest phil. infl. on ⟨him⟩ came from Leibniz.' Gödel's books on quantum mechanics now clearly prove that he did indeed study at least Leibniz' Monadology[3] as early as 1935 (or even earlier). Side by side

---

[1] Henceforth called QM1 and QM2, respectively. Kurt Gödel Papers, Box 6b, Folder 78, item accessions 030106 and 030107, on deposit with the Manuscripts Division, Department of Rare Books and Special Collections, Princeton University Library. Used with permission of Institute for Advanced Study. Unpublished Copyright Institute for Advanced Study. All rights reserved.

[2] Kurt Gödel Papers, Box 6a, Folder 59, item accession 030082.

[3] Item 217 in QM1 also mentions Leibniz' dissertation 'De principio individui.' It reads: 'Der Raum verletzt nach Leibniz das Princ. Id. und daher etwas Ideales. Raum = Princ. Individuationis'.

T. Lethen (✉)
University of Helsinki, Department of Philosophy, History and Art Studies, Helsinki, Finnland
E-mail: tim.lethen@gmx.de

© Der/die Autor(en), exklusiv lizenziert durch Springer-Verlag GmbH, DE, ein Teil von
Springer Nature 2021
O. Passon und C. Benzmüller (Hrsg.), *Wider den Reduktionismus,*
https://doi.org/10.1007/978-3-662-63187-4_3

with rather technical considerations, the books also contain many philosophically orientated comments, in many cases closely connected to Leibniz' Monadology.[4] The aim of this article now is to present and discuss Gödel's interpretation and application of a monadology as it appears in his three books on quantum mechanics.

The following section will concentrate on an analogy between the *universal set* and an objective *thing-in-itself.* Here, Gödel describes what might be seen as a kind of monadologic type theory. Section 3.3 then concentrates on a more complex link between monads and worldviews, with the latter being organized in tree-like structures, representing models of branching time. Section 3.4 briefly reviews the history of branching time, comparing Gödel's notion to those introduced significantly later by Saul Kripke (in connection with A.N. Prior's tense logic) and Nuel Belnap (in connection with relativity and indeterminism).

Throughout the paper, the transcriptions of Gödel's Gabelsberger notes are presented as close to the original as possible. [Square brackets] and (parentheses) are Gödel's own, further additions by the present author are marked in ⟨angle brackets⟩. Although the items in the later Aflenz book are often more elaborate, the particular case always decides, which comment suits the situation best. Also note that the last item of QM2, which has been considered in the Aflenz book, is item 318. As the order of Gödel's notes (and sometimes even within the notes) does not necessarily reflect his overall train of thought, this paper is aiming at reconstructing and presenting Gödel's main ideas. For brevity, we will concentrate on the central passages of the items in question.

3.2   Monads and Types

The application of Leibniz' monadology to quantum mechanics and physics in general starts with a first vague comment on metaphysical systems regarded as 'frames for physical theories.' In item 266 of QM2 Gödel writes:

> Die metaphysischen Systeme sind nichts anderes als verschiedene Rahmen für physikalische Theorien. [Wie sieht das Platonische aus?] Bisher wurde immer nur das Demokrit'sche System als Rahmen in der wirklichen Physik verwendet. In der modernen Physik wird es anders werden. Vielleicht das Leibniz'sche an seine Stelle?

> The metaphysical systems are nothing else but different frameworks for physical theories. [What does the Platonic one look like?] So far only the Democritean system has been used as a framework in real physics. It will be different in modern physics. Maybe the Leibnizian one in its place?

As we shall see, the Leibnizian system in question clearly is his monadology, offering a neat possibility to approximate an objective physical reality and closely linking

---

[4]Other philosophical notes, described in (Kanckos and Lethen 2019), concentrate on neovitalistic concepts.

monads with the idea of different worldviews. This connection between monads and worldviews is describes in item 250 of QM2.

> Vielleicht ist ⟨die⟩ Leibniz'sche Monadolgie eine der Zwischenstufen zwischen Solips. und objektiver Theorie. Charakter: Man hat nicht "ein" "wahres" Weltbild, sondern ebenso viele verschiedene Weltbilder, als es Monaden gibt.
>
> Perhaps the Leibnizian monadology is one of the intermediate stages between solipsism and objective theory. Character: One does not have "one" "true" world view, but as many different world views as there are monads.

The close connection between monads and worldviews is of course no stranger to the Monadology. Here Leibniz states[5] (§57): 'And as the same town, looked at from various sides, appears quite different and becomes as it were numerous in aspects *[perspectivement]*; even so, as a result of the infinite number of simple substances, it is as if there were so many different universes, which, nevertheless are nothing but aspects *[perspectives]* of a single universe, according to the special point of view of each Monad.'

In the aforementionend item 250, Gödel then begins his detailed description of an approximation of an objective theory by pointing at an analogy between the *universal set* and an objective *thing-in-itself.* Whereas type theory can be seen as an approximation of the (non-existent) contradictory universal set, an infinite layering of monadologic structures is able to approximate an objective thing-in-itself as well as an (non-existing) objective physical theory. In his Aflenz book, item 250 begins with the following table:

Analogie zwischen "Allmenge" und "Ding-an-sich":

Russell Antin	*Planck Antin*
In keiner mathematischen Theorie kommt die Allmenge vor. (Nur in der widerspruchsvollen alten.)	In keiner Annäherung an die Wirklichkeit kommt ein vollkommen "objektives" Ding-an-sich vor. (Außer in der widerspruchsvollen klassischen.)

Analogy between "universal set" and "thing-in-itself":

Russell's antinomy	*Planck's antinomy*
The universal set occurs in no mathematical theory. (Only in the contradictory old one.)	The fully "objective" thing-in-itself occurs in no approximation to reality. (Except in the contradictory classical one).

---

[5]Throughout this paper, we use Robert Latta's 1898 English translation of the Monadology.

The role of a theory of types is maybe best described in the corresponding item in QM2:

> Hauptproblem: Für die physikalische Theorie dasselbe leisten, was die Typentheorie für die Logik geleistet hat. D. h., welches ist die <u>Struktur</u> derjenigen[6] Folge von Theorien, welche die klassische Theorie einer "objektiven Welt" zu ersetzen und zu approxim. hat? Ein wesentlicher Bestandteil dabei muss offenbar die "atomistische" (monadologische) Struktur der Welt ⟨sein⟩.

> Main problem: To achieve the same for the physical theory, as type theory has achieved for logic. I.e., what is the <u>structure</u> of the[7] sequence of theories, which has to replace and to approximate the classical theory of an "objective world"? Obviously, the "atomistic" (monadologic) structure of the world has to be a crucial component.

Before we have a closer look at the layers of typed monadologies, it is worth considering Gödel's concept of solipsism. As described in item 250, monadology is seen as an intermediate stage between solipsism and an objective theory. In this respect, solipsism represents an isolated and highly subjective worldview, which does not accept—or rather ignores—the existence of any other viewpoints. The way to overcome this isolation (and at the same time the step into a monadologic world) is the process of an approval of the existence of other entities, i.e. of other monads. In item 254 of the Aflenz book, Gödel very briefly mentions this idea as:

> Monadologie = Solips. + Anerkennung des Du
>
> monadology = solipsism + acceptance of the you

Here, the main constituent is indeed the 'acceptance' or 'approval' ('Anerkennung') of other monads, i.e. of other worldviews, which do exist in different (coexisting) solipsistic worlds. This change of perspective is clearly expressed in item 265 of QM2:

> Quantenmechanik: Einfachste Beschreibung der Welt ist nicht mehr als objektive Welt der Dinge (invariante Beschreibung), sondern Beschreibung der verschiedenen Solips. Welten + Transform. Gesetze.

> Quantum mechanics: The simplest description of the world is no longer as an objective world of things (invariant description), but a description of the different solipsistic worlds + transformation laws.

Once an isolated solipsism has been overcome, one ends up with a first layer consisting of an infinite set of monads. This very first layer can be characterized by the absense of the Leibnizian concept of a *'universal harmony'* (§59), which interconnects the monads. Nevertheless, the simple substances do comprise an inner activity, as Leibniz describes in §18: 'All simple substances [...] have a certain self-sufficiency which make them the sources of their internal activities and, so to speak,

---

[6]Gödel's footnote: transfiniten.

[7]Gödel's footnote: transfinite.

incorporeal automata.' A first level of *harmony* is given by Gödel again in item 250 of QM2:

⟨Die⟩ Leibniz'sche Monadologie ist als Zwischending zwischen Solips. und objektiver Welt nicht willkürlich, denn: ⟨Der⟩ Grund, weswegen man bei Solips. nicht stehen bleiben kann, ist das Du. D. h., es bestehen gleichberechtigte "Selbste", zwischen denen dann offenbar Abhängigkeiten bestehen müssen. [Das ist das Schema (Struktur) der Monadentheorie.] (Die Art dieser Abhängigkeiten kann am besten beschrieben werden durch Spiegelung, aber nicht Spiegelung des Universums, sondern bloß Spiegelung seiner selbst durch tausende von Spiegeln.)

Leibniz's monadology as an in-between between solipsism and the objective world is not arbitrary, because: The reason why one cannot stop at solipsism is the you. I.e., there are equal "selves" between whom there must obviously be dependencies. [This is the schema (structure) of the theory of the monads.] (The nature of these dependencies can best be described as mirroring, but not mirroring of the universe, but merely mirroring of itself through thousands of mirrors.)

Here Gödel follows exactly Leibniz' description of a *universal harmony* amongst the infinite set of monads or simple substances, at least as far as the mirroring relation *between* the monads is concerned. In §56 of the Monadology Leibniz writes: 'Now this connexion or adaptation of all created things to each and of each to all, means that each simple substance has relations which express all the others, and, consequently, that it is a perpetual mirror of the universe.'

Approximating an objective physical theory by means of a kind of monadologic type theory, Gödel certainly has to reject Leibniz' 'consequence', a mirroring of the entire universe. Also, the actual mirroring is an iterated process and takes place in distinct layers of monadologic theories, thus paralleling the layers of set theoretic types. In item 250b of the Aflenz book, Gödel describes this iteration as follows:

Zum Prinzip, welches von T zu T' führt: T_0, es sind nur die Monaden da (ohne Inhalt, d. h. Vorstellungen). T_1, jede Monade enthält das Bild der übrigen (leeren) Monaden. T_2, jede Monade enthält das Bild der übrigen Monaden mit den Vorstellungen aus T_1, usw.

Regarding the principle, which leads from T to T': T_0, only the monads are there (without content, i.e. ideas). T_1, each monad contains the image of the other (empty) monads. T_2, each monad contains the image of the other monads with the ideas from T_1, etc.

At this point, the similarity to set-theoretic types becomes evident: The empty monads resemble 'empty' objects, for instance natural numbers. On the next layer the mirroring monads resemble sets of naturals numbers, and on the following layer they resemble sets of sets of natural numbers, and so on.

Summing up, Gödel clearly underlines a strong parallel between a logical and a monadologic type theory, both constituting an infinite approximation of something unreal and contradictory, the universal set in one case, an objective thing-in-itself in the other. In both cases, a necessary component is an initial step from a single isolated entity to an infinite set of objects. The overall situation is nicely summarized by Gödel himself in item 248 of his Aflenz notes:

Bei konsequenter Durchführung der pos. Quantenmechanik muss irgendwo das intersubjektive Moment hineinkommen (Loskommen vom Solipsismus). D.h., die unendlich vielen Gesichtspunkte (subjektive Weltbilder) sind gesetzmäßig verknüpft, ohne dass diese gesetzmäßige Verknüpftheit auf ein gemeinsames "Ding-an-sich" zurückgeführt wird[8] (= Unterschied gegenüber alter Physik). Hier transfinites Moment, da die Abbildung selbst wieder abgebildet wird.

When consistently carrying out pos. quantum mechanics, the intersubjective element has to enter somewhere (getting rid of solipsism). I.e., the infinitely many points of view (subjective worldviews) are lawfully interwoven, without this lawful interwovenness being traced to a common "thing-in-itself" (= difference from old physics). Here a transfinite element, because the mapping itself is mapped again.

3.3 Branching Time and A Theory of the The Soul

In the model just described, each Monad comprises exactly one point of view, thus establishing a clear one–to–one relation between monads and viewpoints. In a subsequent step, Gödel emphasises a relation between a special kind of Monad and a tree–like structure of viewpoints, the 'person.' His basic idea is again based on the Monadology. In §19, Leibniz writes:

> If we are to give the name of Soul to everything which has perceptions and desires [*appetits*] in the general sense which I have explained, then all simple substances or created Monads might be called souls; but as feeling [*le sentiment*] is something more than bare perception, I think it right that the general name of Monads or Entelechies should suffice for simple substances which have perception only, and that the name of Souls should be given only to those in which perception is more distinct, *and is accompanied by memory* ⟨emphasis added⟩.

And he continues in §26: '*Memory provides the soul with a kind of consecutiveness* ⟨emphasis added⟩, which resembles [*imite*] reason, but which is to be distinguished from it.' Gödel himself states in item 273 (QM2):

> Bewußtsein ist nur möglich durch Mneme.
>
> Consciousness is only possible through memory.

And it is interesting to note that he repeatedly askes the question if a very simple and primitive form of life, namely the *paramecium,* does indeed have a memory.[9]

In a subsequent step, Gödel now connects the concepts *memory* and *viewpoint.* In item 322 of QM2, he obviously describes a mathematical structure $\mathfrak{V} = (\mathcal{V}, \mathcal{P}, \mu, \to, d)$, where $\mathcal{V}$ is the infinite set of *viewpoints,* and $\mathcal{P} \subset \mathcal{V}$ is the set of so-called *possible viewpoints.* If $\mathcal{M}$ is the set of monads, $\mu : \mathcal{V} \to \mathcal{M}$ maps each viewpoint

---

[8]oder Ding-an-sich <u>sehr</u> eigenschaftsarm.

[9]In item 263 (QM2), repeated in the Aflenz book, Gödel writes: 'Behauptung, dass auch ein *Paramäzium* ⟨sic⟩ ⟨ein⟩ Gedächtnis habe! (Nat. 1934, Bleuler, nachsehen Semon).' And in a crossed out item 274 (QM2) he askes: 'Ist das Gedächtnis des Paramäz. ⟨sic⟩ wahr?'.

into the set of monads. $\to$ is a partial order on the set $\mathcal{V}$ of viewpoints, stipulating that $A \to B$ implies $\mu(A) = \mu(B)$, interpreted as 'linked by memory,' both viewpoints A and B belonging to the same Monad. d denotes a metric on $\mathcal{V}$, interpreted as the 'distance' between two viewpoints. Gödel himself describes this structure $\mathfrak{V}$ in item 322 (QM2) as follows:

> Eine "pos." Theorie der Wirklichkeit hat wahrscheinlich folgende Struktur: Grundelemente: die unendlich vielen "Gesichtspunkte". Diese zerfallen in wirkliche Gesichtspunkte und mögliche Gesichtspunkte. [Die ersten sind solche, in denen sich tatsächlich eine Monade befindet.]

> I Zwischen den verschiedenen wirklichen Gesichtspunkten bestehen Beziehungen.
>   a) Die Gesichtspunkte A und B sind [in der Richtung $A \to B$] durch Erinnerung verbunden, d. h. insbesondere, sie gehören derselben Monade an.
>   b) Die Gesichtspunkte A, B sind nah bzw. sind fern voneinander. [D. h., das Bild, welches die Welt für sie bietet, ist mehr oder weniger ähnlich, etc.]

> A "pos." theory of reality probably has the following structure: Basic elements: the infinitely many "viewpoints." These can be devided into real viewpoints and possible viewpoints. [The first ones are those that actually contain a monad.]

> I There are relations between the various real viewpoints.
>   a) The viewpoints A and B are connected [in the direction $A \to B$] by memory, i.e. in particular, they belong to the same monad.
>   b) The points of view A, B are near or far from each other. [I.e., the image the world offers for them is more or less similar, etc.]

Reading on, it becomes clear that the aforementioned basic structure $\mathfrak{V}$ obviously has to be enhanced by a system F of functions which—for each viewpoint—produce an 'image of the world,' as Gödel calls it. Finally, these images are connected by physical laws if their underlying viewpoints are related by the partial order $\to$, the smaller the distance d between two viewpoints, the more similar their images.

> II Jedem Gesichtspunkt ist zwischengeordnet ein "Bild der Welt", dargestellt durch ein gewisses Funktionensystem.
> III Die Axiome der Physik sagen aus, dass zwischen zu verschiedenen Gesichtspunkten gehörigen "Bildern" verschiedene Beziehungen bestehen, falls diese Gesichtspunkte in bestimmter Weise durch Relationen I verknüpft sind.

> II Each viewpoint is mapped onto an "image of the world," represented by a certain system of functions.
> III The axioms of physics say that there are different relations between "images" belonging to different viewpoints, if these viewpoints are linked in a certain way by relations I.

The very last item of QM2 (340) now explains Gödel's concept of a 'person' and interconnects it with viewpoints, monads, the notion of free will, and the phenomenon of time. Under the heading 'Bedeutung des freien Willens und Möglichkeit seiner widerspruchsfreien Vereinigung mit Det.' ('Meaning of free will and possibility of its consistent unification with determinism'), Gödel writes:

Zunächst ist eine Person nur eine Menge von Gesichtspunkten. Wieso bekommt diese Menge eine Struktur (Ordnung)? Hängt zusammen mit Anm. 2, dass ein Gesichtspunkt A irgendwie als "Objekt" eines anderen Gesichtspunktes auftreten kann: B später als A, wenn A Objekt von B, aber nicht umgekehrt.

Das Phänomen der Zeit besteht

1. darin, dass eine mögliche Person nicht ein Gesichtspunkt, sondern eine Menge von solchen ist,
2. dass die Relation des "Objekt-Seins" asym. ist, oder wenigstens besteht eine Person nur aus solchen Gesichtspunkten, für welche das zutrifft, und zwar aus einer möglichst großen Menge. Dies ⟨ist⟩ übrigens vielleicht nur Charakteristikum derjenigen Personen, welche unserer Beobachtung zugänglich sind. [Vielleicht gibt es andere Existenzformen, z. B. mit zweidimensionaler Zeit, etc.]

First of all, a person is only a set of viewpoints. Why does this set get a structure (order)? Related to Note 2, that a viewpoint A can somehow appear as an "object" of another viewpoint: B later than A, if A object of B, but not vice versa.

The phenomenon of time consists in the fact

1. that a possible person is not one viewpoint but a set of such,
2. that the relation of "being an object" is asymmetrical, or at least a person consists only of those aspects to which this applies, and this being a set as large as possible. This is, by the way, perhaps only characteristic of those persons who are accessible to our observation. [Perhaps there are other forms of existence, for example with two-dimensional time, etc.]

Earlier in the same item, Gödel describes his idea of the possibility of a person's influence on the natural flow of time within a branching scheme of viewpoints.

Das Fortschreiten der Zeit besteht darin, dass sich der Gesichtspunkt ändert. Der Gesichtspunkt ist teilweise von außen (Schicksal) bestimmt, teilweise (in sehr geringem Grad) durch meinen Willen. Insofern er durch Schicksal bestimmt ist, haftet ihm das Attribut der "Rätselhaftigkeit und des Geheimnisvollen" an. Durch Training ⟨ist⟩ eine größere Abhängigkeit vom Willen erreichbar (Fakirismus). [Schon durch Technik: leichtere örtliche Verlegung des Gesichtspunktes.] Im Allgemeinen ist die Folge der möglichen Gesichtspunkte [zwischen denen ich zu wählen habe] durch ein Verzweigungsschema in der Richtung wachsender Zeit gegeben. Erinnerung könnte aufgefasst werden als eine Möglichkeit, den Gesichtspunkt aus diesem Schema heraus in die Vergangenheit zu verlegen.

The progression of time consists in the fact that the viewpoint changes. The viewpoint is partly determined from outside (fate), partly (to a very small degree) by my will. Insofar as it is determined by fate, it has the attribute of "enigma and mystery." Through training a greater dependence on the will is attainable (fakirism). [Already through technology: easier relocation of the viewpoint.] In general, the sequence of possible viewpoints [between which

I have to choose] is given by a branching scheme in the direction of increasing time. Memory could be understood as a possibility to shift the viewpoint from this scheme into the past.

Figure 3.1 reflects the overall situation, focusing on a set of viewpoints belonging to the same Monad.

Finally, the difference between 'real' ('wirklichen') and 'possible' ('möglichen') viewpoints is explained—at least to some degree—by sharpening the notion of 'person.' Again in item 340 (QM2), we get the following piece of information:

> Beim Begriff einer Person ist zu unterscheiden zwischen "möglicher" Person und "wirklicher" Person". Nicht je zwei mögliche Personen "passen zusammen". Die Menge der wirklichen Personen muss eine Menge von "zueinander passenden" Personen sein. [Dies ist das Wechselwirkungsproblem der Teilchen.] D. h., die Frage, ob eine Menge eine mögliche Person ist, ist so zu verstehen, ob sie ergänzt werden kann zu einer vollständigen möglichen Menge von Personen. Je nach dem, ob die Ergänzung auf viele oder wenige Arten möglich ist, wird man von einer Wahrscheinlichkeit sprechen.

> In the concept of a person, a distinction has to be made between "possible" person and "real person." Not every two possible persons "match together." The set of real persons must be a set of "matching" persons. [This is the interaction problem of the particles.] That is, the question of whether a set is a possible person is to be understood as whether it can be extended to a complete possible set of persons. Depending on whether the addition is possible in many or few ways, one will speak of a probability.

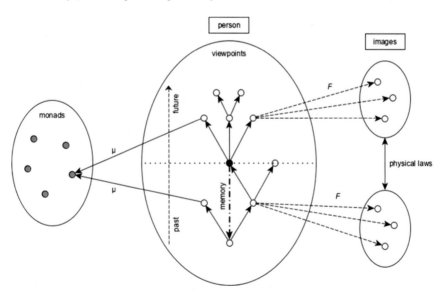

Fig. 3.1 Gödel's theory of the soul in a nutshell. Each person comprises a tree-like structure of viewpoints which all belong to the same Monad. This very Monad rests in one of the viewpoints (marked in black) right between past and future. Memory is regarded as the possibility to change the Monad's viewpoint into the past. Each viewpoint has an 'image of the world,' and if two viewpoints are transitively linked by the relation →, the corresponding images are connected by physical laws. (Not pictured is the metric d.)

3.4   The History of Branching Time Revisited

The crucial role of branching time in Gödel's model suggest a brief revision of its history. In 1950, Arthur N. Prior (1914–1969), well known for his founding work in modern temporal logic, had become interested in the logical studies of the Megarian logician Diodorus. Based on Diodorus ideas concerning time and possibility, Prior represented propositions as infinite sequences of truth values, reflecting the present as well as the future development of that very proposition in terms of truth. If p now is a proposition in this sense, the proposition $\Diamond p$ (again an infinite sequence) has to be interpreted as follows: An element of $\Diamond p$ is true, if an element either at the corresponding or any 'later' position in p is true. Prior (1957) erroneously specified **S4** to be the underlying logical system. Very soon, Saul Kripke, who had discovered the error,[10] wrote a letter to Prior which is described in Øhrstrøm et al. (2010) as 'one of the most important events in the history of logic during the 20th century.' Kripke writes:

> Now in an indetermined system, we perhaps should not regard time as a linear series, as you have done. Given the present moment, there are several possibilities for what the next moment may be like—and for each possible next moment, there are several possibilities for the next moment after that. Thus the situation takes the form, not of a linear sequence, but of a 'tree'. (Kripke 1958)

At the same time, Kripke was able to show that these models of branching time indeed satisfy the logic **S4**. Following Øhrstrøm et al. (2010), this was the very first appearance of the idea of branching time models.

Many years later, in 1992, the American philosopher and logician Nuel Belnap, who had been in close contact to Prior in the 1950s, carried Kripke's idea over to physics, turning *branching time* into a relativistic spatio-temporal variant, *branching space-time (BST),* as an attempt 'to do metaphysics in a mathematically rigorous way, with the desideratum: be compatible with current physical theories' (Placek and Belnap 2012). Whereas traditional physicists regarded the flow of time as a linear temporal order on the infinite set of Euclidean three-dimensional spaces, Belnap now described a tree-like 'causal order' on the Minkowski space of four-dimensional 'possible point events,' thus merging indeterminism and relativity. As a central disadvantage, Belnap (1992) regarded his theory as 'remote from real physics' and later added that it had '(for better or worse) no concept of laws of nature, although it is laws-friendly, since it has modalities and propositions, both rigorously defined' (Placek and Belnap 2012).

Unknown until very recently, Kurt Gödel must have had his idea of branching time already in or even before 1935, immediately integrating it into his theory of the soul. Although Leibniz conception of the soul as a Monad with a memory could have been realized incorporating linear time, Gödel combined it with his thoughts about free will and indeterminism, notions Belnap considered nearly sixty years later. It

---

[10]Kripke had noted that $\Box \Diamond p \vee \Box \Diamond \neg p$ was valid in Prior's system but could not be deduced in **S4**.

should therefore be well worth repeating Gödel's headline for item 340 of QM2, in which he presents the concept of branching time: 'Bedeutung des freien Willens und Möglichkeit seiner widerspruchsfreien Vereinigung mit Det.' ('Meaning of free will and possibility of its consistent unification with determinism.')

The major differences between Gödel's and Belnap's conceptions are twofold: First, Gödel's primitives are viewpoints and a relation consisting of 'links by memory,' whereas Belnap's corresponding primitives are possible point events and a causal ordering relation, respectively. The second difference concerns the underlying topology. Whereas in Belnap's theory the entire set of point events is connected by the causal order, Gödel regards a non-trivial clustering of viewpoints, with certain tree-like clusters, the 'persons,' playing a crucial role, their viewpoints being connected to the same Monad. Furthermore, Gödel's firm connection between viewpoints and physical images allows for a smooth transition from metaphysical structures to physics itself.

3.5 Conclusion

When Oscar Morgenstern noted in his diary (August 28, 1970), that Gödel had stated that his ontological proof for the existence of God was nothing but a logical investigation, he certainly did not see the entire heavy-weighted background. Gödel worked on his ontological proof at least from 1940 on and kept revising it until 1970, when he finally stated that he was satisfied with his latest variant.[11] The motivation for this kind of metaphysical work is precisely expressed in his notebook *MaxPhil IX*[12] (p. 78), where Gödel calls it his main goal in philosophy to define and axiomatize the uppermost philosophical concepts and deduce conclusions and theorems with mathematical rigidity. And in a conversation with Rudolf Carnap in 1940, published in Gierer (1997), Gödel says: 'Man könnte ⟨ein⟩ exaktes Postulatensystem aufstellen mit solchen Begriffen, die gewöhnlich für metaphysisch gehalten werden: "Gott", "Seele", "Ideen". Wenn das exakt gemacht würde, wäre nichts dagegen einzuwenden.' (One could set up an exact system of postulates with such notions that are usually considered metaphysical: "God", "soul", "ideas". If this was done exactly, there would be no objection.) Carnap himself then mentions the analogy to theoretical physics. All these facts clearly underline the essence of Gödel's metaphysical program.

The now discovered and transcribed notes concerning the formalization and axiomatisation of the concept of 'soul' on the basis of Leibniz' Monadology therefore contribute to a deeper understanding of Gödel's metaphysical plan, about which Hao Wang (1987, p. 29) wrote: 'But G appears to have sometimes aimed for at an even higher level of achievement: doing for metaphysics what Newton did for physics ⟨...⟩.

---

[11] For the 1970 variant, see Gödel (1970). For an exposition of Gödel's different variants and their development, see Kanckos and Lethen (2019).

[12] Kurt Gödel Papers, Box 6b, Folder 69, item accession 030095.

I am not able to determine how far G progressed toward such ambitious goals or what evidence he had for believing them attainable.' The notes also show that, although Gödel never published anything about his idea of branching time, he foresaw all the main features which branching space-time brought into physics, nearly sixty years before this theory was finally published.

Acknowledgements The research for this article is a part of the GODELIANA project led by Jan von Plato in Helsinki, Finland. This project has received funding from the European Research Council (ERC) under the European Union's Horizon 2020 research and innovation programme (grant agreement No. 787758) and from the Academy of Finland (Decision No. 318066). I would like to thank Jan von Plato, Maria Hämeen-Anttila, and Annika Kanckos for many fruitful discussions about Gödel's work and his writings. I also would like to thank Marcia Tucker at the Institute for Advanced Study for her generous support.

References

Belnap, N. (1992). Branching space-time. *Synthese, 92*(3), 385–434.

Gierer, A. (1997). Gödel meets Carnap: A prototypical discourse on science and religion. *Zygon, 32*(2), 207–217.

Gödel, K. (1970). Ontological proof. In *Printed in Gödel* (1995) (pp. 403–404). New York: Oxford University Press.

Gödel, K. (1995). *Collected works: Vol. III. Unpublished essays and lectures*, S. Feferman, J. W. Dawson, jr., W. Goldfarb, C. Parsons, & R. Solovay (Eds.). New York: Oxford University Press.

Kanckos, A., & T. Lethen (2019). The development of Gödel's ontological proof. *The Review of Symbolic Logic, 13*, 1–19.

Kripke, S. (1958). *Letter to A. N. Prior (September 3, 1958)*. The Prior Collection, Bodleian Library. Oxford. Printed in (Øhrstrøm, Schärfe, and Ploug, 2010).

Øhrstrøm, P., Schärfe, H., & Ploug T. (2010). Branching time as a conceptual structure. In M. Croitoru, S. Ferré, & D. Lukose (Eds.), *Conceptual structures: From information to intelligence* (pp. 125–138). Cham: Springer.

Placek, T., & Belnap, N. (2012). Indeterminism is a modal notion: Branching spacetimes and Earman's pruning. *Synthese, 187*(2), 441–469.

Prior, A. N. (1957). *Time and modality*. Oxford: Clarendon Press.

Wang, H. (1987). *Reflections on Kurt Gödel*. Cambridge: MIT Press.

The Limits of Reductionism: Thought, Life, and Reality

Jesse M. Mulder

What is the best question reductionists would have to answer but cannot, and why exactly is there no reductionist answer to that question? To answer this question, we need to identify the relevant question. Let us call the question we are looking for the Question.

An obvious candidate for this Question is this one: *what is thought?*—Why? Well, reductionism presents itself as a *thesis* we might come to endorse (or not). If thought is irreducible, then the reductionist does not merely face a bullet that she is unwilling to bite. The reductionist project rests on endorsing a thesis; endorsing a thesis is irreducible; and so the bullet is lethal.

However, this might seem to saddle us with a dualist picture: there is this unique, irreducible part of reality—the part inhabited by beings engaged in the activity of thinking—but *for the rest* reductionism is fine. The existence of rational creatures will then be an annoying detail that spoils the reductionist fun. A reductionist, then, might be tempted to counter with an optimistic promissory note: perhaps some unforeseen future scientific discovery will enlighten us, putting us in reductionist heaven after all.

Hence, I will attempt to push the limits of reductionism further by suggesting as a candidate for our Question: *what is life?* If the reductionist faces an unresolvable problem *here,* squarely within the realm of the natural sciences, she seems to be in more serious trouble. And I will endeavor to claim that, indeed, the reductionist faces such serious trouble. This results not in a dualist picture, but rather in a pluralist one: we must grant *sui generis* status to inanimate nature, life, thought, and perhaps to other realms as well.

Now, a hard-nosed reductionist might resort to an instrumental understanding of biology. She might even adopt an eliminative stance towards life. But we can push

J. M. Mulder (✉)
Utrecht University, Department of Philosophy and Religious Studies, Utrecht, Netherlands
E-mail: j.m.mulder@uu.nl

© Der/die Autor(en), exklusiv lizenziert durch Springer-Verlag GmbH, DE, ein Teil von Springer Nature 2021
O. Passon und C. Benzmüller (Hrsg.), *Wider den Reduktionismus,*
https://doi.org/10.1007/978-3-662-63187-4_4

the limits of reductionism even further, and I will do so by suggesting a third and ultimate candidate for our Question: *what is reality?*

The progression I make with these candidate questions, *thought—life—reality,* at first sight is one of expansion. We first focus on a very limited domain (the thinkers), then widen our scope towards a larger domain (life) and end up with the largest possible domain (reality). I will boldly suggest, however, that the last question, on reality is, in fact, the very same as the first question, the one on thought. Thus rounding the circle, we find ourselves not with a dualist or pluralist but rather with a monist picture (but, of course, not of a reductive variety).

Before presenting these three candidate Questions, I will briefly introduce my conception of reductionism, in Sect. 4.1 below. Then follow the promised three candidate Questions: in Sect. 4.2 *what is thought?;* in Sect. 4.3 *what is life?;* in Sect. 4.4 *what is reality?* I conclude in Sect. 4.5, where I elaborate briefly on the 'transformative' motivation behind my choice for these three candidate Questions.

4.1   Reductionism

In principle, many different kinds of view could justifiably be named reductionist. A Berkeleyan idealist reduces everything material to perceptions; a classical atomist reduces everything to indivisible atoms and their motions within the void. And there are local reductionisms as well: one might take the social realm to be reducible to interactions among individuals,[1] and yet accept those individuals as irreducible to, say, the level of biology or physics. Or one might take life to be reducible to mere material interactions, and yet take *conscious* life to have an irreducible status of its own (being impressed, perhaps, by Chalmers' 'hard problem').[2]

Here, I will take issue with what is plausibly the most popular form of global reductionism: physicalism. Roughly put, physicalism is the view that (future, or idealized) physics will tell us what the fundamental elements of reality are and how they behave—and that this is all there is. "All else supervenes on that", to borrow David Lewis's sweeping phrase.[3]

Difficult questions immediately emerge: how should we cash out this slogan, that "all else supervenes" on the physical facts? Different versions of physicalism are on offer in this regard, yet I will not discuss any one of them in detail. Nor will I critically question whether supervenience is the proper notion to be used here (though it is, indeed, a problematic notion).[4] These questions I will leave to those invested in the

---

[1] See, for instance, Bratman (2014) for an attempt to understand group agency in terms of individual agency, and Rödl (2014a) for a non-reductive stance.

[2] See Chalmers (1995)—I briefly return to Chalmers' 'easy' and 'hard' problems shortly.

[3] See Lewis (1986, pp. ix–x). The phrase comes from his formulation of the view he calls 'Humean supervenience'.

[4] For instance, supervenience claims do no more than state that supervening properties vary with their subven-ing property base (very roughly). They do not explain *why* that is so. And this is considered by most to be unsatisfactory. See, e.g., Kim (1998).

reductionist project. I do not intend to target some specific theoretical rendering of reductionism, but rather the broader reductionist *picture* that is holding us captive.[5] A picture, indeed, which, despite their efforts, holds captive even some of the most ardent enemies of reductionism (we will come back to this in Sect. 4.5).

Let me evoke a more vivid image of what this reductionist picture involves by presenting the following typical quote, taken from the introductory chapter of a book that presents an overview over the contemporary consciousness debate:

> [E]verything happening in the universe is ultimately a process involving the basic forces of nuclear attraction, electromagnetism, and gravity, in various combinations. Digestion is a process by which food is broken down into usable energy for the body. This is a chemical process: complex starches, say, are converted into the glucose our cells need to power their activities. And the chemistry is explainable in terms of more basic atomic interactions: various attractions and repulsions at the atomic level make up chemical reactions. There's nothing else to them in the final analysis. (Weisberg 2014, p. 13).

Here it is suggested that the entities and activities one finds in animate nature—organisms, digestion, etc.—can be reduced to the "various attractions and repulsions at the atomic level", i.e., at the level of physics. A typical reductionist claim.[6]

Now, I do not want to suggest that what Weisberg writes concerning digestion is false. Of course, the digestion of complex starches *can* be explained on the level of chemistry and even physics. We are not missing anything relevant on those levels once we've found a sufficiently detailed account. There are no further, hidden, quasi-physical factors involved. And this suggests that nothing more can be said *tout court*. But compare: a full description can be given of my copy of Weisberg's book in terms of patterns of black ink on white paper. Nothing needs to be added to that description either—there are no additional hidden quasi-patterns, printed with invisible ink by means of some ghostly, intensional printer. Yet the book, in a different sense, *does* contain more: it contains an overview over the consciousness debate. Likewise, Weisberg's own quote points to what the digestion of complex starches involves beyond the mentioned attractions and repulsions at the atomic level: his quote speaks of *food,* which is to be *transformed into useable energy* for some *body.* These specifically biological, teleological, concepts get lost once we phrase the whole story in terms of attractions and repulsions. Just like the actual contents of the book get lost once we describe it in terms of ink-patterns.

The reductionist will now claim that, since *obviously* there is nothing *in addition* to the attractions and repulsions (or patterns of ink), these further observations about this instance of digestion (or about my copy of Weisberg's book) must in some way be reducible. In some way: contemporary physicalists typically resist committing themselves to any specific form of reduction. For perhaps it is impossible to give an

---

[5]I mean to be echoing Wittgenstein here, who remarks in his Philosophical Investigations, regarding his own earlier *Tractatus* view, that "a picture held us captive" (Wittgenstein 1953, §15).

[6]Indeed, Weisberg expresses a disarming enthusiasm towards this claim later in his book: "For my part, I think it's super amazing that we might 'just' be a physical system. I find it incredibly inspirational to think of myself and the rest of humanity in this way." (Weisberg 2014, p. 46).

exhaustive account of the contents of Weisberg's book, or of digestion as a biological process, in terms of attractions and repulsions. This is why many physicalists nowadays call their view 'non-reductive physicalism' (which, by the way, does still belong under the label of reductionism as I will understand it here).[7]

This presentation of the reductionist position already suggests the diagnosis I have in mind: by restricting our attention to just the physical level, we lose sight of the very phenomenon we were studying. It disintegrates under our very eyes. The various physico-chemical interactions that make up a given digestive process belong together as a process of digestion. Digestion is the reason why they are occurring *in this order and sequence;* it is even the reason why they are occurring *at all.* But digestion, here, is not an additional physical quantity or force. Similarly, the various ink patterns in my copy of Weisberg's book *belong together* as expressions of the content he intended to convey in writing it. That content is the reason why they are there. The very coherence of those chemical processes (and of these ink patterns), their unity as a process of digestion (or as a book) thus depends on something that is simply *not to be found* on the physical level. We don't realize what went thus missing, because we tacitly project that unity onto the physical level. To borrow an apt metaphor from Crawford Elder: the reductionist tacitly relies on the *shadows* that higher-level entities (an organism, a contentful book) cast onto the physical level, without realizing that those shadows are dependent on that which casts them (cf. Elder 2011).

The reductionist is likely to object. "Of course, not everything is explained by the *immediately present* physical goings-on! But that doesn't mean that there is no reductive explanation of, say, digestion, or the content that a book conveys. We just have to look at the wider context!" And the reductionist is surely right—accordingly, this essay doesn't end here.

Notice, now, that the gesture towards an adequate reductive story is so far not much more than a promise. If you are empirically minded, you might claim an easy victory at this point: "The burden of proof lies with the reductionists; let us simply wait and see how far they get with their attempts at keeping their promise!" And this is surely fair: even in the most fundamental and paradigmatic cases, reductive theses have tended to be untenable upon closer inspection.[8]

Such an attitude, however, forever leaves it open that the reductionist project might eventually be vindicated. On this route, then, one will not be able to find a suitable

---

[7]Nonreductive physicalism is a curious phenomenon. It brings to light that, indeed, reductive physicalism is a picture holding those defending it captive: it holds them captive to such an extent that it is alright for them to grant it a self-cancelling name ('nonreductive' physicalism).—Another symptom of this situation is the following. In the face of the many difficulties that physicalists encounter in their attempts to formulate their distinctive, physicalist thesis, Alyssa Ney has come to defend physicalism as an "attitude": "physicalism is an attitude one takes to form one's ontology completely and solely according to what physics says exists" (Ney 2008, p. 9).

[8]Color is a nice example; see Stroud (2000) for an excellent in-depth discussion. The issue whether classical genetics reduces to molecular biology is another fine instance—see, for an overview, Brigandt and Love (2017), and see also Sect. 4.3 below.

candidate Question—the requirement is, after all, that we show why the reductionist *cannot* answer it, not merely that she *hasn't as of yet* answered it, or is not likely to find a conclusive answer.

A squarely philosophical approach, on the other hand, will attempt to investigate the very viability of such a reductionist project. Is life, or contentfulness, amenable to reductionist treatment *at all?* It is in this latter spirit that I will be discussing the very ideas of thought, life, and reality in what follows.

4.2   What is Thought?

Nowadays, the opinion is widespread that, while qualitative or phenomenal consciousness constitutes a serious or 'hard' problem, thought doesn't. Indeed, Chalmers famously grouped many of the issues related to thought together under the label "easy problems", which "seem directly susceptible to the standard methods of cognitive science, whereby a phenomenon is explained in terms of computational or neural mechanisms" (Chalmers 1995, p. 201).

Under the heading of "easy problems", Chalmers here summarizes a mechanistic conception of the rational mind, which Gödel has sought to refute until the end of his days. Gödel saw in his own incompleteness theorem a first step towards that refutation, yet, with typical caution, he usually did not venture beyond the following disjunctive conclusion:

> My incompleteness theorem makes it likely that mind is not mechanical, or else mind cannot understand its own mechanism. (Gödel, as quoted in Wang (1997, p. 186).

Although Gödel did, in addition, express allegiance to Hilbert's "rationalistic optimism", which would eliminate the second option, he did not want to rest content with such a mere conviction (Wang 1997, p. 185 f.).

If Chalmers' quote captures the received view, the prospects for Gödel's project may look dim. Yet if we read Sebastian Rödl's following claim, this pessimistic estimate, and Chalmers's labeling of thought as an "easy problem", looks to be entirely misplaced:

> Perhaps it is sensible to dream of some development of natural science by which sensory consciousness comes to be within its reach. However, this dream is obviously incoherent in the case of [thought]. (Rödl 2014b, p. 492)

Why does Rödl think it so obvious that thought will never come within the reach of natural science? And how could this be of help to Gödel?

We can unpack Rödl's bold claim by going back to the simple observation that Lewis Carroll famously captured in his parable of Achilles and the tortoise over a century ago (Carroll 2005). Suppose you (validly) infer C from A and B. What does this involve? You think A and B, and then you think C. But that is not enough: just *adding* the judgment C is no inference. Rather, in making this inference, you are aware *that C follows from A and B*. So perhaps what is missing is this extra premise: $A \& B \to C$. However, the addition doesn't help: now we imagine that you judge A, B, and $A \& B \to C$. And then you proceed to judge C. This juxtaposition of judgments again does not amount to seeing that C *follows* from the previous three.—The lesson to draw is that *no* addition will do the job.

Now suppose that through some "computational or neural mechanism" I add a judgment, C, to given judgments A and B. This isn't inference, since my arrival at C doesn't rest on a recognition that C follows from A and B. To be sure: the mechanism might be logically sound, i.e., such that it only produces representations that *in fact* follow from the given ones. That still doesn't make it inference, for this is not reflected in the representations themselves. Being asked why I judge C, I can only say: "I don't know." If my judgments spring from a mechanism of the mentioned kind, I will simply find myself saddled with them, in utter incomprehension.

An alternative suggestion springs to mind. Let us say that the mechanism is more sophisticated, in that it provides me with an additional judgment about those three: "I judge C because it follows from my judgments A and B". This suggestion also misfires, and it is instructive to see why: it separates my awareness of the validity of my inference from that very inference. It attempts to portray my (first-order) inference as something that is independent of my (second-order) awareness of its validity. However, if the inference itself doesn't involve that very validity, it simply isn't inference—and no amount of *additional* representations, be they second-order or not, will turn something that isn't an inference into an inference. (This was Carroll's insight.)

We can make another interesting observation here: the proposed second-order judgment, taken at face value, *already includes the entire inference.* That is to say, not if it is something I find myself saddled with, in utter incomprehension. But if it is taken to signify my coming to see that, indeed, C follows from A and B, then this simply is my concluding C on the basis of A and B (given that I have already judged A and B).[9]

In short, then, making an inference is not something *separate* from being conscious of making it.[10] The unity of judgments in my inference *is* my consciousness of that unity. Rödl puts this as follows: inference is *self-conscious,* where the term "self-consciousness" signifies not a consciousness of some object, a 'self', but rather this

---

[9]Is it really? What if I have reason to doubt C, won't I then rather reject A or B, or suspend my judgment?—Sure. This is, however, no objection to the point under consideration; it rather illustrates that point. My judgments aren't separate elements lying around in my awareness, unconnected. They are united, and their union is precisely my consciousness of their union. Famously, Kripke (1982) presented a reading of Wittgenstein on following a rule ('Kripkenstein') that does take note of problems such as the one concerning inference briefly outlined here. Kripke, however, fails to realize that the solution lies in the recognition of the self-consciousness of thought. That is why he ends up with his 'skeptical solution' (Kripke 1982, p. 66 ff.).

[10]Perhaps there are such things as unconscious inferences. That is no objection to what I say here. At most, it is the mere observation that an account of such unconscious inferences is still lacking. See also Nagel (2012), esp. Chap. 4, and Kitcher (2011), esp. Chap. 15, §4.

peculiar phenomenon of something being its own comprehension; see, e.g., Rödl (2018, Chap. 1).[11] In every inference, I know that I am inferring. Put differently: inference knows itself to be inference.

It requires only a little reflection to see that the same holds for the unity comprehended *within* one judgment: that of predication. Within the framework of the present essay, this brief gesture towards a generalization of our conclusions with regard to inference will have to suffice to introduce the claim that thought, as such, is self-conscious (in Rödl's sense). Thought knows itself to be thought.

Now, science in general is concerned with comprehending empirically given phenomena.[12] So the phenomena science studies are *by definition* independent from the comprehension sought. And cognitive science indeed attempts to approach cognition in this way: there is this empirically given phenomenon, cognition, and we are now trying to understand it, preferably by identifying the relevant "computational or neuronal mechanisms" underlying it. It thus seeks to *add* an understanding to something that does not by itself already include its own understanding. And this fundamentally misrepresents the object of study.

This should suffice to see the point of Rödl's remark, quoted above, that the sort of understanding cognitive science seems to be after is "obviously incoherent": its object is a self-conscious phenomenon, i.e., a phenomenon that includes its own comprehension, yet it strives to provide an account of that phenomenon from the outside.[13]

The following objection is bound to arise: cannot these two projects peacefully coexist? There is the *first-personal* comprehension of thought, which may very well be 'self-conscious' in Rödl's sense, and there is a scientific, *third-personal* comprehension of thought, which grounds it in something that is not first-personally comprehended as its ground.

However, as Rödl observes elsewhere (Rödl 2018, §4.3), this suggestion treats the two types of comprehension of thought as two "perspectives" on thought. And this presupposes that that on which they are perspectives is as it is independently of being comprehended from either perspective. And this, as we saw, is not the case. The self-consciousness of thought is not something *additional:* it is thought.

---

[11]The Anscombean tradition in contemporary philosophy of action makes the same point with regard to intentional action: my intentional action is not something *independent* of my knowledge of it. Anscombe expresses this, for instance, by saying that this knowledge, practical knowledge, is "the cause of what it understands". See Anscombe (1957, §48), and also Rödl (2007, Chap. 2).

[12]How do we know this to be the case, concerning science? Have we discovered it by empirical investigation? Obviously not; compare Rödl (2018, p. 16). The statement that science is concerned with comprehending given phenomena merely conveys the comprehension that is included in science. Science is self-conscious as well.—We will return to this in Sect. 4.4.

[13]Notice that I do not thereby claim that cognitive science is a doomed project in its entirety. There might be very many processes that can be described using the 'automatic' sort of 'cognitive systems' one finds in cognitive science (early visual processing is a good example). But not thought.

The self-consciousness of thought is an insight that meshes rather interestingly with Gödel's above-mentioned search for a satisfactory argument against a mechanical conception of mind. Consider, again, his statement:

> My incompleteness theorem makes it likely that mind is not mechanical, or else mind cannot understand its own mechanism. (Gödel, as quoted in Wang 1997, p. 186).

The self-consciousness of thought provides what Gödel was looking for in his search for a sufficient ground to exclude the second option mentioned in this quote. For, the notion of a (rational) mind that cannot understand itself (or "its own mechanism") *just* is the impossible notion of a non-self-conscious yet thinking mind.[14]

To conclude, then, the answer to the question *What is thought?* at least involves the insight that thought is self-conscious. The reductionist insists that everything in the end reduces to physical goings-on. Physical goings-on, however, are not self-conscious: physical goings-on do not include their own comprehension. Hence, there can be no answer to the question *What is thought?* that will be satisfactory for the reductionist. Moreover, since the reductionist must accept that there is such a thing as science, she must accept that there is such a thing as thought. In other words: this question is a fine candidate for our Question.

4.3   What is Life?

Thought, then, poses a problem for the reductionist. But, she might say, isn't the presence of thinking beings just a contingency? And do not thinking beings make up only a tiny portion of the universe? A humbler reductionist victory might still be possible with respect to the vast realm of non-thinking things. And perhaps, then, the irreducibility of thought, its self-consciousness, can be considered an anomaly of sorts, a local fluke in the cosmic reductive order of things.

A more thorough rejection of the reductionist program, then, will have to show even this humbler reductionist victory to be impossible. We can arrive at such a more

---

[14]The question what Gödel's incompleteness theorems imply with regard to the philosophy of mind is, of course, a vexed one. See, for instance, the two essays by Putnam and Penrose on the topic in Baaz et al. (2011, Chaps. 15 and 16). For what it is worth: I do not think there is any route *from* Gödel's incompleteness theorems *to* conclusions about thought. For one thing (and I owe this observation to Albert Visser), the tendency to traverse that route often results in a dispute over the claim whether the human mind is 'more powerful' than anything a mechanism could do (see the mentioned essays by Putnam and Penrose), where the notion of 'power' involved equivocates between the very abstract, theoretical concept of what lies within the 'power' of a given formal system on the one hand, and concrete mental abilities on the other.

Instead, as I suggested here, the interesting project would be to try to comprehend the significance of Gödel's results in the light of a proper understanding of thought, and that is, in the light of thought's self-consciousness.

thorough rejection by reflecting on the idea of *life*—inspired by Michael Thompson's masterful discussion thereof.[15]

What is life? Unlike thought, living beings are natural, given, non-self-conscious objects, and thus objects fit for scientific inquiry in the sense mentioned above. Hence it is natural to start answering the question what life is by composing a list of features that distinguishes life from non-life. One feature frequently associated with life is organization, or complexity. Entropy would be a suitable, physical measure of complexity.[16] And living things are indeed physical systems displaying remarkably low entropy values. But, even if in fact *all and only* living things display entropy values below a certain determinate threshold, this doesn't tell us much: it is not physically *impossible* that arbitrarily low entropy states are occasionally reached by non-living systems. And given that the entropy scale is continuous, what makes this specific entropy value so special? Does the difference between a certain living cow at one time and its fresh corpse a second later really consist in just the insignificant increase in entropy (if there even is such an increase)?

The situation is different if, instead of entropy, a thicker notion of complexity is invoked: living things are organized in the sense that they are composed of *organs: 'organ-ized'* (cf. Thompson 2008, p. 38). This makes sense; living things are indeed composed of parts in a rather idiosyncratic sense. Living parts, organs, are, for instance, not detachable: as Aristotle was wont to say, a detached hand is a hand only 'homonymously'; see, e.g., Aristotle (1998, 1036b30–32). Another way of putting this is to say that in living things, the whole comes before the parts. Indeed, the whole *makes* its parts—which is beautifully illustrated by our detailed knowledge of embryological development. Compare, say, a car: cars can be assembled by putting together prefabricated parts, so that the whole comes after the parts.—In any case, once we ask what exactly distinguishes organs from non-living parts, it becomes clear that we have gone round in a circle: organs are parts of, specifically, *living* things.

It is observations such as these that prompt Thompson to conclude that "every candidate list-occupant must strike the sub-metaphysical Scylla of ['entropy'] or else sink into the tautological Charybdis of 'organs'". (Thompson 2008, p. 39).[17]

Isn't Thompson jumping to conclusions here? Perhaps organization *by itself* is insufficient to define life, but would not the situation be different if *more* features were added? In biology textbooks, one typically finds mentioned in this context such

---

[15] See especially Thompson (2008, Part I). For a much more detailed exposition and extension of the views Thompson puts forward, see Mulder (2016).

[16] I ignore the fascinating question whether entropy itself can be reductively accounted for in terms of statistical mechanics. See, for a classical treatment, Sklar (1993, Chap. 9).

[17] In this quotation, Thompson originally has 'DNA' instead of 'entropy'. The point is in the end the same; con-sidering the hypothetical situation in which all and only living things turn out to contain DNA, Thompson writes: "The judgment about DNA, if it were true, would only show how resource-poor the physical world really is. It could make no contribution to the exposition of the concept of life [...]" (Thompson 2008, p. 37).

features as metabolism, growth, adaptation, response to stimuli, etc. Shouldn't we focus on cases in which *all* of these are present?

Now, to be sure, Thompson does consider several of those, but the point is not merely to question whether these features, individually or jointly, single out all and only the living things. Instead, reflections of the kind just rehearsed are meant to illustrate that when it comes to life, we should not be interested in a list of features that happens to be extensionally adequate; what we should want to comprehend are rather the typical categories in terms of which we understand living things—the 'vital categories', *organ* for instance. And these categories turn out to resist being captured in merely physico-chemical terms. That is why, for each and every 'list-occupant' we find two options: either we understand that list-occupant in a merely physical way, in which case we can always ask what *that* has to do with life (this is what Thompson calls 'sub-metaphysical'), or we understand it (implicitly or explicitly) in a way that presupposes the very concept of life (what Thompson calls 'tautological').

Organ is, thus, the specifically 'vital' version of the abstract notion of *parthood*. Likewise, one can find 'vital' analogues of other basic categories. An instructive further example is the vital analogue of process, which Thompson calls "life-process". Consider the following illustrative quote:

> In a description of photosynthesis, for example, we read of one chemical process [...] followed by another, and then another. Having read along a bit with mounting en-thusiasm, we can ask: "And what happens next?" If we are stuck with chemical and physical categories, the only answer will be: "Well, it depends on whether an H-bomb goes off, or the temperature plummets toward absolute zero, or it all falls into a vat of sulfuric acid [...]" That a certain enzyme will appear and split the latest chemical product into two is just one among many possibilities. Physics and chemistry, ade-quately developed, can tell you what happens in any of these circumstances—in any circumstance—but it seems that they cannot attach any sense to a question "What happens next?" sans phrase. (Thompson 2008, p. 41)

But there *are* answers to such 'what happens next'-questions—biology is full of examples. This illustrates that life-processes, processes for which Thompson's 'what happens next'-question makes sense, exist for a *reason,* a reason that is not to be found by looking at their merely physico-chemical components or phases (recall Weisberg's observation, quoted in Sect. 4.1 above, that digestion consists in the conversion of ingested food, such as "complex starches", in order to power the activities of "our cells").

To see what this 'reason' really consists in, first note that even the very notion of existence takes on a specific shape in the case of life—"to be, for living things, is to live" (Thompson 2008, p. 27). And indeed, as philosopher of biology John Dupré never tires to point out: "a static cell is a dead cell" (Dupré 2013, p. 30).[18] To live is to be actively engaged in life-processes that mutually sustain and enable each other.

---

[18]Dupré's work is full of examples that illustrate the distinctiveness of life. He doesn't take these observations all the way to a decidedly philosophical articulation of their ground—and therefore he ends with a rather generic insistence that we should shift to a 'process metaphysics' across the metaphysical board. See Dupré (2012, 2013, 2018).

Life-processes are always embedded in, and thus unified by, the whole life cycle of which they form part. This life cycle is the full expression of the *life form*—the 'vital' counterpart of the more generic category of a *natural kind.* The life form, then, is the 'reason' to which life-processes inevitably point. In traditional vocabulary: life is everywhere teleological, not in the external sense of serving some *further* purpose or aim, but in the internal sense of being its own end. Living beings differentiate themselves into mutually supporting parts, and their life cycle differentiates into mutually supporting life-processes, and these differentiations everywhere serve the purpose of realizing the life form in question.[19]

In any case, we have again merely scratched the surface of a huge topic; much more needs to be said on these vital categories and their relation to contemporary issues in (the philosophy of) biology, such as the status of biological species, evolutionary theory, etc.[20] Nevertheless, I hope I have said enough to motivate my proposal to put up the question *What is life?* as a candidate Question. If the above is roughly correct, that question cannot be answered without using the vital categories, which is precisely what the reductionist would have to do. It is, thus, another fine candidate for our Question.

Now, the reductionist might resort to an instrumentalist understanding of life, such as the one advocated by Alexander Rosenberg (1994). In brief, this comes down to claiming that even though the vital categories are indispensable, they are merely useful instruments that do not capture what is really "out there". Metaphysically speaking, this amounts to an *elimination* of life. We only think there is life because *for us* the vital categories are indispensable. (Although one might wonder *for whom* these categories are precisely indispensable—aren't we ourselves alive, too?)

This indicates that we might try to push the limits of reductionism even further. Let us proceed to our final attempt.

---

[19]It is interesting, in this light, to read in one of Gödel's recently transcribed notebooks the following 'philo-sophical remark' on life:

> Life is obviously an imperfect structure, which therefore attracts matter from outside [...] and takes it up into its structure. The new structure obviously acts upon itself with a "destructive force", resulting in the emission of urea and carbonic acid. Does this entire process result in a perfection of the original structure? (Our body only deteriorates over time and only our minds get better.) All of this shows, that life, continuously perfecting itself, comes from something that has no perfection. (Crocco et al. 2017, p. 7, my translation)

It does not seem to occur to Gödel here that destruction may be an integral and crucial part of what it is to live. In any case, Gödel obviously granted life a sui generis position within his philosophical thought—as is also evi-dent from this quote: "Life force is a primitive element of the universe and it obeys certain laws of action." (Wang 1997, p. 193). See also Kovac (218, §2.2.6) for discussion. For discussion of these and related issues see Mulder (2016).

[20]For discussion of these and related issues see Mulder (2016).

4.4   What is Reality?

For the endgame, let us return to the quote from Weisberg provided in Sect. 4.1. Ultimately, Weisberg says, everything boils down to physical interactions involving the basic forces of electromagnetism, gravity, and the nuclear forces. (Or, as my father-in-law likes to say: ultimately, life is nothing but moving stuff around.) So, a reductionist could say, ultimately, our conclusions concerning thought and life do not matter. Perhaps there is no way of understanding those phenomena in a satisfactorily reductive manner. Perhaps utter confusion and incomprehension is somehow inevitable when it comes to such complex matters. Still, the fundamental insight (says the reductionist) that it all boils down to those physical interactions stands.

Here we must enter the lion's den and challenge the reductionist on the very level she takes to constitute the fundament of reality. At first sight, this may seem to be an absurd strategy. Surely, we cannot point to any reductionistically troublesome concepts, or phenomena, *on the physical level*—isn't the reductionist supposed to *accept* precisely all and only those physical concepts as being, so to speak, metaphysically serious?

That is surely right, but there is another way of challenging the reductionist on this level. The reductionist assumes as a matter of course that these physical concepts (or the perfected versions thereof a future completed physics will discover) can be *isolated* from the concepts and forms of explanation discussed earlier. It is quite obvious, so the reductionist thinks, that reality might just as well have harbored *only* physical stuff, and no life or thought *at all*. After all, the occurrence of life, and of thought, are contingent happenings, which might just as well not have happened.

My aim in pointing this out is not to challenge the possibility of a physical world without living or thinking beings. Rather, I want to point out that that possibility does not in itself suffice to underwrite the sort of isolation of the physical sphere that the reductionist presupposes.

There are various ways in which we could challenge that isolation. One way would be to argue that the physical is *as such* already also material for the living, so that the very idea of life must be accounted for anyway—whether life 'materializes' or not. That, however, would yield only a deepening of the candidate Question posed in the previous section, and hence no new candidate Question.

Instead, consider again Weisberg's claim that "it all boils down to physical interactions". What does the "all" there signify? It stands for *everything,* or for *reality.* Reality is exhausted by those physical interactions. Is this concept, the concept of reality, a physical concept? No. It isn't even an empirical concept. We didn't discover, empirically, that there is this thing, "reality". Rather, the very act of discovering something, the very idea of empirical science, rests on that concept. The concept of reality is an *a priori* concept.

If this is right, the reductionist cannot point to anything physical in order to answer the question *"what is reality?"*. One way in which this has been recognized in the literature is as follows. Summing up all the physical happenings, we sum up all of reality, according to the physicalist. But having summed up all of those happenings, it is still open whether or not there are *more* physical facts than those mentioned, and thus we need to add an extra, non-physical fact: "That's all".[21]

However, here the "that's all" component is still conceived as an addition, something disjoint from the physical facts themselves. Yet with every physical statement we make, we assert that what we say to be the case *is so;* we put it within the domain that is supposed to be circumscribed by the "that's all" addition. In Wittgenstein's words: "When we say, and mean, that such-and-such is the case, we—and our meaning—do not stop anywhere short of the fact; but we mean: this-is-so." (Wittgenstein 1953, PI §95). McDowell famously paraphrased Wittgenstein's thought as follows: "When one thinks truly, what one thinks *is* what is the case." (McDowell 1996, p. 27). And this idea, Wittgenstein's *this-is-so,* McDowell's being-the-case, already is the idea of reality as a whole. Rödl writes:

> [T]he concept of what is, the concept of a fact, the concept of something real, does not signify a part, an aspect, a limited region of—of what? yes: of—what is, the facts, reality. The concept of what is is not a concept of anything limited [...]. It is not con-tained in anything larger than it. (Rödl 2018, p. 56)

What the reductionist must accept as fundamental—a conception of reality in physical terms—everywhere rests on this concept of reality, of being the case, of "what is" (the Greeks discussed it under the heading of *being*). How do we account for this idea? Not by means of physics, we saw. Nor can we account for it by means of *any* form of empirical science. Especially, we cannot redirect the question to cognitive scientists, expecting that they come up with a psychological, evolutionary, or otherwise empirical explanation of the idea of reality.

This brings us back to our topic in Sect. 4.2: there, I claimed that thought, because of its self-consciousness, cannot be the topic of *any* empirical science. Now we see that the same holds for reality. And this is no coincidence, for reality is just another name for the self-consciousness of thought.[22] Whenever we say (or think), and *mean,* that something is the case, we know ourselves to mean just that. Thought knows that its ultimate object is reality; reality is the ultimate object of thought. This we know in every particular thought that we have, just as we know ourselves to be thinking in every particular thought that we have. McDowell added to the above quote: "So since the world is everything that is the case [...] there is no gap between thought, as such, and the world." (McDowell 1996, p. 27). Perhaps we can read Gödel's following 'Philosophical Remark' in the light of the coincidence of the idea of reality with the limitlessness of thought:

> Aristotle's proof that the intellect is not corporeal and has no bodily organ at all [...] is after all the antinomical character of the 'all'.*

---

[21]Chalmers (2012) includes such "totality truths" as facts of a separate kind in his Carnap-inspired *Constructing the World.*

[22]This insight, that the idea of reality is nothing other than the self-consciousness of thought, is central to Rödl's (2018) *Self-Consciousness and Objectivity*—its title can be read accordingly.

* Or better, the possibility to make 'all' in turn an object and to transcend beyond that (the unboundedness [*Uferlosigkeit*]).[23]

In any case, it is no coincidence that both Rödl and Gödel find inspiration in Aristotle's fascinating argument to the effect that thought (the intellect) is immaterial.[24]

4.5   Concluding Remarks on the Question

It was not my aim to argue at full length for the challenge that each of the three candidate Questions poses for reductionism. That would require much more space and focus than I have allowed myself in this essay. I chose to discuss three different questions, and hence had to make do with a rather sketchy treatment of each of them, for two reasons. First, I wanted to indicate the radically different levels at which one might fundamentally challenge the reductionist. Thus understood, the three candidate Questions are relatively independent from each other. But secondly, and more importantly, the order in which I presented the candidate questions, the progression from the one to the other, was meant to bring out three consecutive layers at which one might free oneself from the prevailing reductionist picture. By way of conclusion, I will attempt to explicate this 'transformative' aim.

For those with anti-reductionist leanings, our self-understanding as human beings often provides a good entry point for an argument against reductionism. The attitude here is one of retreat: reductionism has conquered realm after realm—the starry skies, the realm of the living, etc.—but luckily there remains this final anti-reductionist stronghold: the human being, ourselves. This is reflected in the traditional opposition between the sciences and the humanities.[25] The first question thus centered on thought, conceived as the center of this stronghold.

As I indicated, this stance does constitute an effective resistance to reductionism, but it simply buys into the whole reductionist world view outside of the realm of human affairs. It is, in essence, a dualist picture. We can overcome this dualism by resisting the reductionism not only when it concerns us, human beings, but also within its own realm. To this end, I turned to *life* as the topic of my second candidate Question. Here we resist the temptation to retreat that defines the first stance. We squarely oppose the reductive assimilation of the realm of the living; we discover

---

[23]This remark can be found in Gödel's unpublished manuscript *Max Phil VI*, 404. I take the quote from Engelen (2016, p. 172), who also provided this English translation. (The Aristotelian proof Gödel here alludes to can be found in his *De Anima III* (Aristotle 1984, 429a18–29).) Of course, as the *-footnote to this remark indicates, Gödel connects this Aristotelian insight with issues in set theory he was thinking about—Russell's antinomy, the unlimited expansion of the set universe, the idea of proper classes. See Engelen (2016) for discussion.

[24]Rödl (2014b, §2) discusses this argument extensively with the help of Plato's version thereof in his *Theaetetus*.

[25]Here, the traditional opposition between *Erklären* and *Verstehen* of course comes to mind.

that the dualist position we were faced with was forced upon us merely because of the aggressive reductionist expansion.

From this second stance, then, we do not end up with a dualist picture, but rather with a pluralist one. Animate nature does not reduce to inanimate nature; thought does not reduce to life; and perhaps there are more levels to be discerned.[26] Yet, an asymmetry remains that continues to support reductive thinking: physics still appears to be *fundamental*. Life depends on matter, even thought appears to require a healthy brain. The various levels thus strike us as optional extras that the physical realm could also do without. Hence my final candidate Question: *what is reality?* Reflection on this question may disarm the reductive spell that makes it seem obvious that the physical level is fundamental. We come to see that this is a mistake: rather, the notion of reality is fundamental. This notion is involved in all of the various levels, so that we may say that this third stance constitutes a progression from pluralism to a non-reductive monism. Reductive monism isolates one level, typically the physical one, to the exclusion of others. It distorts our comprehension of reality. Non-reductive monism, by contrast, does not seek to ground its monism empirically, in a specific, favored part or aspect of reality. It grounds itself, rather, in the self-consciousness of thought. On the face of it, this is a formulation of what Rödl calls, following Hegel, *absolute idealism:* "reason is the certainty of consciousness of being all reality" (Rödl 2018, p. 15).

Freeing ourselves stepwise from the reductive picture that held us captive, we thus find ourselves with absolute idealism. Traversing this transformative path, we recognize that the three candidate Questions are really one.

Acknowledgements This work was supported by the Dutch National Science Foundation NWO, VENI Grant (grant number 275-20-068). I am grateful to Albert Visser for valuable feedback on an early version.

References

Anscombe, G. (1957). *Intention*. Oxford: Wiley & Blackwell.
Aristotle. (1984). *On the soul (Trans. J.A. Smith)*. Princeton: Princeton University Press.
Aristotle. (1998). *Metaphysics (Trans. H. Lawson-Tancred)*. London: Penguin.
Baaz, M., Papadimitrou, C. H., Putnam, H. W., Scott, D. S., & Harper, C. L. (2011). *Kurt Gödel and the foundation of mathematics*. New York: Cambridge University Press.
Bratman, M. (2014). *Shared agency*. Oxford: Oxford University Press.
Brigandt, I. & Love, A. (2017). Reductionism in biology. In E. N. Zalta (Ed.), *The Stanford encyclopedia of philosophy (Spring 2017 Edition)*.
Carroll, S. B. (2005). *The new science of evo devo—Endless forms most beautiful*. New York: WW Norton.

---

[26]This form of pluralism can be found in the work of philosophers belonging to the so-called 'Stanford School' in philosophy of science. See, e.g., Dupré (1993, 2018), and Galison and Stump (1996).

Chalmers, D. J. (1995). Facing up to the problem of consciousness. *Journal of Consciousness Studies, 2,* 200–219.

Chalmers, D. J. (2012). *Constructing the world.* New York: Oxford University Press.

Crocco, G., Van Atten, M., Cantu, P., & Engelen, E.-M. (2017). Kurt Gödel Maxims and philosophical remarks volume X. Available via HAL Archives Ouvertes: https://hal.archives-ouvertes.fr/hal-01459188. Accessed: 1. Sep. 2020.

Dupré, J. A. (1993). The disorder of things. Cambridge: Harvard University Press.

Dupré, J. A. (2012). *Processes of life: Essays in the philosophy of biology.* Oxford: Oxford University Press.

Dupré, J. A. (2013). Living causes. *Arist Soc P., 87*(1), 19–37.

Dupré, J. A. (2018). A manifesto for a processual philosophy of biology. In D. J. Nicholson & A. Dupré (Eds.), *Everything flows: Towards a processual philosophy of biology.* Oxford: Oxford University Press.

Elder, C. L. (2011). *Familiar objects and their shadows.* Cambridge: Cambridge University Press.

Engelen, E.-M. (2016). What is the link between Aristotle's philosophy of mind, the iterative conception of set, Gödel's incompleteness theorems and god?: About the pleasure and the difficulties of interpreting Kurt Gödel's 'philosophical remarks'. Episteme. In G. Crocco (Ed.), *Kurt Gödel: philosopher-scientist* (pp. 171–188). Aix-en-Provence: Presses Universitaires de Provence.

Galison, P., & Stump, D. J. (Eds.). (1996). *The disunity of science: Boundaries, contexts, and power.* Stanford: Stanford University Press.

Kim, J. (1998). The mind-body problem after fifty years. In A. O'Hear (Ed.), *Current issues in philosophy of mind* (pp. 3–21). Cambridge: Cambridge University Press.

Kitcher, P. (2011). *Kant's thinker.* Oxford: Oxford University Press.

Kovac, S. (2018). On causality as the fundamental concept of Gödel's philosophy. *Synthese, 197,* 1803–1838.

Kripke, S. A. (1982). *Wittgenstein on rules and private language.* Cambridge: Harvard University Press.

Lewis, D. K. (1986). *Philosophical papers* (Vol. II). New York: Oxford University Press.

McDowell, J. (1996). *Mind and world.* Cambridge: Harvard University Press.

Mulder, J. M. (2016). A vital challenge to materialism. *Philosophy, 91*(2), 153–182.

Nagel, T. (2012). *Mind and cosmos.* New York: Oxford University Press.

Ney, A. (2008). Physicalism as an attitude. *Philosophical Studies, 138*(1), 1–15.

Rödl, S. (2007). *Self-consciousness.* Cambridge: Harvard University Press.

Rödl, S. (2014a). Intentional transaction. *Philosophical Explorations, 17*(3), 304–316.

Rödl, S. (2014b). Review of transcendental philosophy and naturalism. *European Journal of Philosophy, 22*(3), 483–504.

Rödl, S. (2018). *Self-consciousness and objectivity.* Cambridge: Harvard University Press.

Rosenberg, A. (1994). *Instrumental biology or the disunity of science.* Chicago: University of Chicago Press.

Sklar, L. (1993). *Physics and chance: Philosophical issues in the foundations of statistical mechanics.* Cambridge: Cambridge University Press.

Stroud, B. (2000). *The quest for reality: Subjectivism & the metaphysics of colour.* New York: Oxford University Press.

Thompson, M. (2008). *Life and action: Elementary structures of practice and practical thought.* Cambridge: Harvard University Press.

Wang, H. (1997). *A logical journey from Gödel to philosophy.* Cambridge: MIT Press.

Weisberg, J. (2014). *Consciousness.* Cambridge: Polity.

Wittgenstein, L. (1953). *Philosophical investigations.* Oxford: Blackwell.

True or Rational? A Problem for a Mind-Body Reductionist

5

Michał Pawłowski

5.1   Introduction

The problem presented in this essay touches upon an issue of reductionism in philosophy of mind, represented by various forms of physicalism. A whole range of reductionist physicalisms (hereafter I use words "physicalism" and "reductionism" interchangeably, except in a passage, where the so called non-reductionist physicalism is discussed briefly) has to answer a question which poses a serious difficulty shaped in the form of a dilemma.

The dilemma goes as follows: a physicalist must accept that either physicalism is false or it lacks rational grounding, since the question "How can be the truth and the rational justification of physicalism compatible?" is unanswerable in a way expected by such a philosopher. As I try to show, these two features: the rational grounding and the truth of this belief exclude each other and as a result a reductionist must face an unsolvable problem concerning the possibility of their simultaneous assertion. Unsolvable, as every time one tries to assert reductionism's truth, it proves to be self-undermining. The argument for this thesis has been provided by various formulations of an objection originating from Karl R. Popper's remarks on materialism (Popper and Eccles 1983) and all together they pose a danger for the identity theory, epiphenomenalism, behaviourism and the mind-body functionalism. A physicalist might try to answer this objection leading them to the conclusion that their belief is either false or irrational (which is an idle task). They may also try to adopt other, non-reductive approaches, yet resulting in their ceasing to a be a reductionist.

In the first part of this essay I present the argument against physicalism and distinguish it from other arguments frequently proposed against it (such as the inverted

M. Pawłowski (✉)
University of Warsaw, Warschau, Poland

qualia argument) as well as from other dilemmas (just to mention now the famous Hempel's dilemma). Then I go on to discuss possible answers of a reductionist and show their irrelevance in some cases and non-reductionist consequences in others. Finally, I sum it up in short conclusions.

5.2   The Dilemma for a Reductionist—The Neopopperian Argument against Physicalism

The dilemma has been partially presented in *The Self and its Brain,* written together with John C. Eccles, where Popper argues that the identity theory is incompatible with rationalism. Among various remarks showing inconsistency of this or other forms of materialism,[1] he quotes John B. Haldane's objection against materialism. It goes as follows:

> [...] if materialism is true, it seems to me that we cannot know that it is true. If my opinions are the result of the chemical processes going on in my brain, they are determined by the laws of chemistry, not of logic. (Haldane as cited in Popper 1983, p. 75).

As Popper points out, the existence of computers makes the argument easily refutable, as their functioning based on laws of physics on no account contradicts functioning according to laws of logic. Later on, however, he goes on to develop Haldane's argument in the form of a dialogue between an interactionist and a physicalist and reaches a conclusion that the physicalist lacks an external point of reference while assessing his beliefs. For the interactionist such an anchor is provided by the objects of the World 3 (such as logic in general), whose existence is denied by the physicalist. Consequently, the physicalist's stance is irreconcilable with rationalism (Popper and Eccles 1983, p. 81); see the discussion in Pawlowski (2019).

Similar objections are proposed in Edward Feser's discussion of Friedrich A. von Hayek's causal functionalism and arguments against it presented by Popper (Feser 2006a). According to Popper, as Feser notices, functionalism is self-undermining, since it ignores a core fact that distinctive properties of human language are its argumentative and descriptive functions (in contrast to the signal one, characterizing for example bees' system of communication): if one reduces a proposition's meaning to its functional role between inputs and outputs of a given (computer or neural) system, a person's having it has nothing to do with a logical inference from other beliefs. And then none of these propositions is rationally justified, thus also a proposition asserting physicalism's truth (Feser 2006a, pp. 307–310). Feser supports this approach with another example: if causal powers of certain beliefs do not depend on their contents or meanings (as they are rooted in electrochemical properties of

---

[1] Such as: "[...] the radical physicalist must adopt radical behaviourism and apply it to himself: his theory, his belief in it, is nothing; only the physical expression in words, and perhaps in arguments— his verbal behaviour and the dispositional states that lead to it—is something." (Popper and Eccles 1983, p. 60)

neural processes), we cannot insist that our thoughts "Socrates is a man" and "all men are mortal" serve as rational justification of the thought "Socrates is mortal" due to their meaning. Was their content different, like "Fido is a dog" (but causal powers the same), nothing would change. Thus no belief is rationally justified, and *a fortiori* also the belief that physicalism is true (Feser 2006b, pp. 150–154).

Mariusz Grygianiec offers a brief reconstruction of the argument against the identity theory in a form of a syllogism (Grygianiec 2016). Analogously, it can be also presented in this framework as an objection against functionalism, which I proposed in (Pawlowski 2019):

1. If functionalism is true, then every mental state (including all beliefs) is identical[2] to a functional state.
2. If every belief is identical to a functional state, then every inference comes down to passing from one functional state of a brain to another, given definite causal connections (laws) bounding inputs and outputs of the system and described by a theory T.
3. If every inference comes down to passing from one functional state of a brain to another, given definite causal connections bounding inputs and outputs of the system and described by a theory T, then these states possess their causal powers only because of their functional properties (not owing to their meanings or contents that they are bound with), ascribed basing on this theory.
4. If these states possess their causal powers only because of their functional properties (not owing to their meanings or contents that they are bound with), ascribed basing on a theory T, then there is no such belief that could serve as a justification for a different one (since only functional connections, bounding inputs and outputs of the system define which conviction is entailed by which).
5. If there is no such belief that could serve as a justification for a different one, then no belief can be rationally justified.
6. If no belief can be rationally justified, then also functionalism cannot be rationally justified.
7. Conclusion: if functionalism is true, then it cannot be rationally justified.

Conclusion: functionalism is false or cannot be rationally justified (Pawlowski 2019).

It is worth noticing that the same applies not only to the identity theory (when the first premise will be the following: *If materialism is true, then every mental state (including all beliefs) is identical to a neurophysiological state* [compare Grygianiec (2016)]), but also to epiphenomenalism, because it explicitly postulates mental states' lack of causal powers (compare Jackson 1982), and to behaviourism as a simpler version of functionalism. As a consequence, a mind-body reductionist has to answer the question, how he or she can believe in the truth and the rational justification of their position at the same time. This, according to that argument, is impossible, since either physicalism is true or cannot be rationally grounded (its truth entails its

---

[2]One does not have to differentiate here between the type identity and the token identity.

irrationality). Although a physicalist might employ various strategies to avoid facing this agonizing conclusion, as will be discussed later on, none of them remains valid.

Another interesting fact to note is that a similar argumentative strategy is taken by the opponents of the (more generally understood) metaphysical naturalism. They argue that it leads to its own self-refutation, just to mention Alvin Plantinga (2002) and Thomas M. Crisp (2016). Analogous objection was also provided by C. S. Lewis, against whom Elizabeth Anscombe famously argued, accusing him of confounding reasons and causes (Anscombe 1981). Here, however, it should be pointed out that the reductionism, being a stronger thesis, claims that mental states are "nothing more and above" than the "so-and-so" (physical states, natural states, etc.). Consequently, the critique need not be referred to that case.

At the same time, this dilemma posing a danger for different forms of physicalism (the identity theory, epiphenomenalism, behaviourism and functionalism) should be distinguished from other arguments against it. For example, similarity to another well-known argument used against physicalism, the Hempel's dilemma, is rather superficial. Hempel's dilemma entails that either physicalism should be defined in terms of contemporary physics and then most presumably is false or in terms of future or ideal physics and then is trivial (Hempel 1969; Stoljar 2017). Similarly, Feser argues that scientism is either false, because its ambition to explain the whole reality only with a use of scientific methods is unattainable, as they must always rest on some prescientific (philosophical) assumptions, or trivial, when under "science" one understands also these assumptions (Feser 2014, pp. 10–13). Here, however, the alternative lies not between this belief being false or trivial, but rather between the falsity and the lack of rational justification. In both cases the latter part of the alternative (triviality/irrationality) includes the truth of physicalism. However, what makes them different are obvious character in the first, and irrationality of such truth in the second case.

Moreover, the dilemma "the falsity or (the truth and) the irrationality" is for sure a way different argumentative strategy against physicalism than most objections proposed in philosophical debate. For the sake of brevity I refer to arguments considering probably the most significant form of physicalism nowadays, functionalism. While the inverted qualia argument and other objections such as the what-is-it-like-to-be argument by Thomas Nagel (1974) or the Mary argument proposed by Frank Jackson (1986) try to show that functionalism fails to explain the whole richness of the reality's aspects, Ned Block's argument proves that it cannot guarantee the possession of consciousness to these, and only these structures to which it should be ascribed (it cannot exclude the "liberalism" and the "chauvinism" at the same time) (Block 1993), and the Kripkean objection against functionalism posed by Joseph Levine (1993) simply attempts to prove its falsity, the here presented dilemma offers a much more interesting solution. Functionalism (and with it some other reductionist approaches) might be true, yet for an undeniably high price: admitting it, ambitions to maintain rational approach must be abandoned. On the other hand, the desire for rationality entails its falsity.

5.3   (The Impossibility of) Physicalist Replies

Confronted with the syllogistic version of the argument, a physicalist might attempt to answer it in many ways. The first solution is to try to challenge the validity of this inference through overruling one of the premises.

It might be argued that the premise 1) is too wide: it should not be spoken about all mental states, but only about some of them. Ned Block, for example, makes a restriction that only "narrow" mental states should be involved, namely those ones, whose truth-conditions can be found only inside a given person (Block 1993). The stipulation is intended to guarantee that the Twin Earth critique of Hillary Putnam (1973) will not be applicable here. As a result of it, the entire inference would collapse. Block offers a counterexample (meant to prove "wide" mental states' irrelevance in this field) according to which my duplicate, created by the exchange of all atoms would be functionally identical to me, yet he would lack some of my memories. In such cases it should be denied that all beliefs are identical to some functional states and thus the premise 1) should be refuted (see Pawlowski 2019).

However, as argued elsewhere (Pawlowski 2019), these two replies, Putnam's counterexample of the Twin Earth and Block's narrow mental states do not have to be accepted: it might be shown that it is simply the case that the world in which water is H_2O offers a different input than in the world in which water is XYZ, while considering identical functional systems. Even if these convictions generate the same outputs, they will be different functional states, since inputs *de facto* differ. It seems that one does not have to "save" the initial assumptions of functionalism from Putnam with such a reply, simply because there are much easier counterarguments. Moreover, the accurate duplicate example is in turn rather untenable, as it is in no case sure, whether the duplicate would be characterized by psychological continuity with me or not. Thus it is also not sure if he really would not have the same beliefs as me. One does not have to accept this reply. For instance, Derek Parfit (1984, p. 239) argues that this is exactly the case that my duplicate would be psychologically continuous with me; see Pawlowski (2019). Consequently, both problematic claims: the one proposed by Putnam and the one defended by Block (which tries to answer the Twin Earth example) seem dubious. At the same time, a functionalist, willing to explain the phenomenon of mind, should not simply ignore some mental states as irrelevant: functionalism must be a general theory of mind or it will not be such a theory at all. Thus, however controversial the premise 1) could seem, it is fully justified and reasonable.

While the premises 5) and 6) are rather uncomplicated tautologies and the premise 2) is an uncontroversial description of the functionalist position (as presented in Block 1993, or Levine 1993), a physicalist might argue that one should not so easily accept the premises 3) and 4) and the requirement of justification and rationality of a belief following from them. They might claim that this requirement is too restrictive, similarly as it is proposed in various discussions with sceptics, who can be criticized for their inflated expectations towards cognitive certainty (see Wittgenstein 1975). Rational justification, a physicalist will say, is exactly this: the fact of a belief's accurate inference from other beliefs, basing on causal powers described by a given theory T with regard to appropriate inputs and outputs. There is nothing more behind it and whoever demanding more from it expects the impossible.

A simple counterexample might be proposed to this objection. If meanings and contents of beliefs play no role, since their rationality and the level of justification are rooted only in causal powers of a given functional system, no mistake would be possible. Was my brain so designed (natively or resulting from a surgery), that all operations "$2 + 2 =$" would lead to the result "5", everything would be perfectly fine. Similarly to previously discussed examples by Feser with Socrates and Fido, we need meanings or contents of mental states to be able to speak about their rational inferences. Without an external point of reference other than the causal powers (also causal powers of the others' brains), such as standards of rationality, we would not have any point of reference. Thus the critique of inflated standards of rationality leads to its understated level.

Being unable to rejoin the argument, a reductive physicalist may adopt another view. They might be inclined either to a weaker or a stronger approach. The weaker is represented by the so called non-reductive physicalism, generally exemplified by views accepting the principle of supervenience—mental states' dependence on physical states as their "base states"—yet without an attempt to reduce the former to the latter. However, one does not have to remind various difficulties non-reductive physicalism encounters, just to mention the causal exclusion problem (see Kim 1998), to spot the irrelevance of this strategy. Adopting non-reductive physicalism, a physicalist simply ceases to be a reductionist. Similarly with a stronger view, that is eliminativism—an attempt to eliminate mental states as objects of the non-scientific common-sense psychology. Whenever a reductive physicalist tries to become an eliminativist and excludes mental states as non-existing, he or she no longer presents a reductive attitude.

Finally, a physicalist might attempt to ignore the results of the dilemma and overrule the principle of truth, correspondingly with Friedrich Nietzsche's remarks from *On Truth and Lies in a Nonmoral Sense:*

> In some remote corner of the universe, poured out and glittering in innumerable solar systems, there once was a star on which clever animals invented knowledge. That was the haughtiest and most mendacious minute of "world history"—yet only a minute. (Nietzsche 1976, p. 42)

Or follow Richard Rorty's liberal ironist's attitude and reconcile with the impossibility of a fully objective vocabulary describing the world (Rorty 1993). Then, they will say, physicalism is only a useful tool for the description of the world. They might do that, but no longer can their beliefs be described as rationally justified.

Consequently, any time a reductive physicalist will try to avoid the consequences of the presented dilemma, he or she will have to cease being a reductionist or a rationalist—that is accept one of the question's results.

5.4 Conclusions

The dilemma, originating from Popper's writings and creatively developed by other authors, poses a great danger for mind-body reductionists. Its conclusion that reductive forms of physicalism force their representatives to admit either the falsity or the lack of rational justification of their beliefs is bounding, because all attempts to face it are idle. The inference is logically correct and based on reliable premises: thus any attempt to ignore its results requires abandoning physicalist reductionism. One option is adopting non-reductive physicalisms (or trying to eliminate the concept of mental states), another—falling into nonrational beliefs.

A physicalist then, confronted with the question "How can you believe in the truth and rational justification of your view at the same time?" is unable to reply it, since the argument-dilemma proves these two features to be incompatible. Any sensible answer leads to a change in the intellectual position and abandoning reductionism.

References

Anscombe, G. E. M. (1981). A reply to Mr CS Lewis's argument that "Naturalism" is self-refuting. In G. E. M. Anscombe (Ed.), *The collected papers of GEM Anscombe II: Metaphysics and the philosophy of mind* (pp. 224–232). Oxford: Blackwell.

Block, N. (1993). Troubles with Functionalism. In A. I. Goldman (Ed.), *Readings in the philosophy of science* (pp. 231–253). Cambridge: MIT Press.

Crisp, T. M. (2016). On naturalistic metaphysics. In K. J. Clark (Ed.), *The Blackwell companion to naturalism* (pp. 61–74). London: Wiley Blackwell.

Feser, E. (2006a). Hayek the cognitive scientist and philosopher of mind. In E. Feser (Ed.), *The Cambridge companion to Hayek* (pp. 287–314). Cambridge University Press: Cambridge.

Feser, E. (2006b). *Philosophy of mind. A beginner's guide.* Oxford: Oneworld Publications.

Feser, E. (2014). *Scholastic metaphysics: A contemporary introduction.* Piscataway: Transaction Books.

Grygianiec, M. (2016). Die popperschen Herausforderungen für den Materialismus. *Logos i Ethos, 2*(42), 103–115.

Hempel, C. (1969). Reduction: Ontological and linguistic facets. In *Essays in honor of Ernest Nagel* (pp. 179–199). New York: St. Martin's Press.

Jackson, F. (1982). Epiphenomenal Qualia. *The Philosophical Quarterly, 32*(127), 127–136.

Jackson, F. (1986). What Mary didn't know. *The Journal of Philosophy, 83*(5), 291–295.

Kim, J. (1998). *Mind in a physical world: An essay on the mind-problem and mental causation.* Cambridge: MIT Press.

Levine, J. (1993). Materialism and qualia: The explanatory gap. *Pacific Philosophical Quarterly, 64*, 354–361.

Nagel, T. (1974). What is it like to be a bat? *The Philosophical Review, 83*(4), 435–450.

Nietzsche, N. (1976). On Truth and Lie in an Extra-Moral Sense (frag.) In *The Portable Nietzsche, sel. and trans. by W. Kaufmann* (pp. 42–47). London: Penguin Books.

Parfit, D. (1984). *Reasons and persons.* Oxford: Oxford University Press.

Pawlowski, M. (2019). Neopopperowski argument przeciw funkcjonalizmowi. *Filozofia Nauki, 3*(27), 77–86.

Plantinga, A. (2002). Introduction. In J. Beilby (Ed.), *Naturalism defeated? Essays on Plantinga's evolutionary argument against naturalism.* Ithaca: Cornell University Press.

Popper, K. R., & Eccles, J. C. (1983). *The self and its brain.* London: Routledge & Kegan Paul.

Putnam, H. (1973). Meaning and reference. *The Journal of Philosophy, 70*(19), 699–711.

Rorty, R. (1993). *Contingency, irony and solidarity*. Cambridge: Cambridge University Press.
Stoljar, D. (2017). Physicalism. In E. N. Zalta (Ed.), *The Stanford encyclopedia of philosophy*. Stanford, CA: Stanford University.
Wittgenstein, L. (1975). On certainty (Ed. by G. E. M. Anscombe & G. H. von Wright, trans. by G. E. M. Anscombe & D. Paul). Oxford: Blackwell.

Teil II
Naturwissenschaftliche Perspektiven

Why Reductionism does not Work

6

George F. R. Ellis

6.1 Emergence, Reductionism, and Causation

Kurt Gödel opposed the reductionist viewpoint of logical positivism (Wang 1997, p. 173). For example he wrote:

> Even if we adopt positivism, it seems to me that the assumption of such entities as concepts is quite [as] legitimate as the assumption of physical objects and that there is quite as much reason to believe in their existence. (Wang 1997, p. 174)

He asked more specifically as regards biology:

> Is there enough specificity in the enzymes to allow for a mechanical interpretation of all functions of the mind? [...] I believe that mechanism in biology is a prejudice of our time which will be disproved. (Wang 1997, p. 192)

The arguments I give below show he is correct in both cases. The reductionist explanation he opposed is doomed to failure.

Life emerges out of physics in a bottom-up way: atoms are made of electrons and protons, molecules of atoms, cells of molecules, physiological systems (including brains) out of cells, and organisms out of physiological systems. The issue is whether higher levels have causal powers, or not: are they just epiphenomena, in view of the alleged causal completeness of the underlying physics?

The thesis of this essay is that, (i) Reductionism does not work because strong emergence occurs in many important cases. In particular in biology, "More is dif-

G. F. R. Ellis (✉)
Mathematics Department, University of Cape Town, Cape Town, South Africa
E-mail: george.ellis@uct.ac.za

© Der/die Autor(en), exklusiv lizenziert durch Springer-Verlag GmbH, DE, ein Teil von
Springer Nature 2021
O. Passon und C. Benzmüller (Hrsg.), *Wider den Reduktionismus*,
https://doi.org/10.1007/978-3-662-63187-4_6

ferent" (Anderson 1972), since the whole is more than just the sum of its parts. (ii) This emergence is possible because downwards causation takes place right down to the lower physical levels, hence arguments from the alleged causal completeness of physics and supervenience are wrong. Emergence of higher level contextually branching dynamics is made possible by downward causation, whereby lower levels, including the underlying physical levels, are conscripted to higher level purposes through time dependent constraints. The higher levels are thereby causally effective, so strong emergence occurs.

No violation of physical laws is implied. The key point is that outcomes of universally applicable generic physical laws depend on the context when applied in specific real world biological situations (Atmanspacher and Graben 2009). The same is true for example in the case of digital computers.

In this essay, I look at the nature of strong emergence, the fact that it occurs, and how it is enabled by downward causation (Sect. 6.2); the ways that downwards causation is possible (Sect. 6.3); physics examples (Sect. 6.4); digital computers as a very clear example (Sect. 6.5); and biology examples, including the brain (Sect. 6.6). I end with a discussion of how top down action causes branching of physics at the lower levels, and hence undermines the argument from supervenience against strong emergence (Sect. 6.7). This makes clear key questions reductionists would have to answer, but cannot.

6.2   Strong Emergence Occurs

In this section, I look at the definition of strong ontological emergence, and its relation to ontology (Sect. 6.2.1); its outcome, the existence of modular hierarchical structures (Sect. 6.2.2), which is the proper context to consider strong emergence (Sect. 6.2.3). It is useful to distinguish different types of strong emergence (Sect. 6.2.4). But does strong emergence occur? I argue that it does (Sect. 6.2.5), and particularly that abstract entities have causal powers (Sect. 6.2.6). They clearly cannot be explained in a reductionist way: they have a completely different nature than physical variables. The crucial point is that it is downward causation that enables strong emergence to occur (Sect. 6.2.7),

6.2.1   Strong Ontological Emergence

Emergence is usually classified firstly into ontological and epistemological emergence, and secondly into strong emergence and weak emergence. This paper is concerned with *strong ontological emergence*, that is, firstly it does not "characterize the concept of emergence strictly in terms of limits on human knowledge of complex systems" (O'Connor and Wong 2015), rather it considers emergence as a phenomenon that exists in its own right (whether or not humans know about it, and independent of whether it concerns issues to do with the mind and brain).

Ontology
My take on ontology is as follows:

(i) Physical objects exist at all scales, so for example a desk exists just as much as the atoms out of which it is made (cf. Eddington 1927), and that is true whether humans know about it or not. In that sense, this agrees with a materialist position;

(ii) Any entity that can be demonstrated in either an experimental or counterfactual way (Menzies 2001) to have a causal effect on physical entities that exist (item i) must also be said to exist, else we will have uncaused events occurring in the physical universe. This leads to the conclusion that for example algorithms, ideas, and social conventions are abstract entities that ontologically exist, as discussed below (Sect. 6.2.6). In that sense this disagrees with a materialist position.

Strong Emergence
Secondly, I follow Chalmers' definition of strong emergence:

> **Strong emergence**: A high-level phenomenon is strongly emergent with respect to a low-level domain when the high-level phenomenon arises from the low-level domain, but truths concerning that phenomenon are not deducible even in principle from truths in the low-level domain. (Chalmers 2006)

This is a clear statement of the principle "More is different" (Anderson 1972). By contrast, Chalmers states as regards weak emergence,

> **Weak emergence**: We can say that a high-level phenomenon is weakly emergent with respect to a low-level domain when the high-level phenomenon arises from the low-level domain, but truths concerning that phenomenon are unexpected given the principles governing the low-level domain [...] It often happens that a high-level phenomenon is unexpected given principles of a low-level domain, but is nevertheless deducible in principle from truths concerning that domain. (Chalmers 2006)

My position, in common with Leggett (1992), is that such an "in principle" deduction is almost always illusory: it cannot in fact be done in a way that depends only on lower level quantities. As stated by Leggett:

> No significant advance in the theory of matter in bulk has ever come about through derivation from microscopic principles. (...) I would confidently argue further that it is in principle and forever impossible to carry out such a derivation. (...) The so-called derivations of the results of solid state physics from microscopic principles alone are almost all bogus, if 'derivation' is meant to have anything like its usual sense. Consider as elementary a principle as Ohm's law. As far as I know, no-one has ever come even remotely within reach of deriving Ohm's law from microscopic principles without a whole host of auxiliary assumptions ('physical approximations'), which one almost certainly would not have thought of making unless one knew in advance the result one wanted to get, (and some of which may be regarded as essentially begging the question).

Essentially the same is stated by Laughlin (1999) in the context of superconductivity, and Scott (1999) in the case of the brain. One must look at the issue in the relevant hierarchical context (see Sect. 6.2.4), and then such emergence at some higher level L_2 from a lower level L_1 almost always in fact depends on concepts and entities at the higher level L_2.

The third possibility is Bedau:

> **Nominal emergence**: The notion of a macro property that is the kind of property that cannot be a micro property, and is not strongly emergent. (Bedau 2002)

Thus this is the kind of emergence in the mind of reductionists. It is the case where more is just the sum of the parts: there are no surprises.

6.2.2   The Outcome: Modular Hierarchical Structures

The outcome of emergent processes is the existence of Modular Hierarchical Structures, with very different kinds of causation occurring at each level of the hierarchy. The way this works out is radically different in the cases of the the natural sciences and human sciences hierarchies (see Table 6.1).

Key to understanding emergence is conceptual clarity as to what variables belong to what levels in this hierarchy. Note that in the natural sciences hierarchy, higher levels correspond to larger scales (and lower energies), and only physical variables come into play. By contrast in the life sciences hierarchy, quite different kinds of variables come into play at higher levels: indeed they include non-physical variables (see Sect. 6.2.6).

Table 6.1 The emergent hierarchy of structure and causation for inanimate matter (left) and life (right) as characterised by academic discipline. Causality of appropriate kind occurs at each level in both cases, described by suitable variables for that level. The bottom four levels are common to both sides (life emerges from ordinary matter)

	Inanimate matter	Living Matter
Level 10	Cosmology	Sociology/Economics/Politics
Level 9	Astronomy, astrophysics	Psychology, Rationality
Level 8	Space & planetary Science	Physiological systems
Level 7	Geology, Earth Science	Cell Biology, Cell signaling networks
Level 6	Materials, Structures	Molecular biology, Supramolecular chemistry
Level 5	Physical chemistry, crystals	Biochemistry
Level 4	Atomic physics: elements	Atomic physics: elements
Level 3	Nuclear physics	Nuclear physics
Level 2	Particle physics	Particle physics
Level 1	Fundamental theory	Fundamental theory

The emergent higher level structures on the life sciences side are *Adaptive Modular Hierarchical Structures*, with specific functions at each level. Key features are:

- Structure: underlies function, and shapes what happens (Campbell and Reece 2005; Mossio et al. 2009)
- Hierarchical: different levels of emergent complexity arise, each with appropriate emergent entities/variables and causation for that level (Campbell and Reece 2005)
- Modular: abstraction and information hiding occur with controlled interfaces, allowing module modification without destroying system function (Booch 2006; Sales-Pardo 2017)
- Networks: interactions between modules form causal networks (Boogerd et al. 2005), with preferred network motifs (Milo et al. 2002; Alon 2006) and perhaps hubs
- Adaptive: at each level entities adapt to the environment for that level, which includes higher levels. This takes place on evolutionary, developmental, and functional timescales.

The Basis of Life
All living systems are of this nature (Campbell and Reece 2005; Rhoades and Pflanzer 1989; Randall et al. 2002), where the structural and functional details have been determined by Darwinian evolutionary processes (Mayr 2002; Campbell and Reece 2005) in the sense of an Extended Evolutionary Synthesis (Pigliucci and Müller 2000; Noble 2008; Carroll 2005).

Artificial Systems
Complex technological systems such as digital computers and aircraft have many of the characteristics of life, also being based in modular hierarchical structures (Simon 1996; Booch 2006), except in general they are not as adaptive.

6.2.3 Strong Emergence (Hierarchical Context)

In looking at what strong emergence happens, we must interpret Chalmers' "truths concerning that phenomenon are not deducible even in principle from truths in the low-level domain" (Chalmers 2006) in the context of this hierarchy (Table 6.1). I claim it should be taken to mean:

> **Strong emergence (hierarchical context)**: Truths concerning a phenomenon at a higher level are not deducible even in principle only using variables defined at a lower level in the hierarchy.

Variables that can only be defined in terms of relations at a higher level (e.g. crystal structure, which is level 5) are thus in this case by definition irreducible to variables defined at a lower level (e.g. the level of electrons and protons, Level 2).

Thus if a reductionist says "Yes but the crystal structure is nothing but an aggregation of electrons and ions", i.e. it is describable at Level 2, the response is that that statement tells you nothing about the specific crystal structure, for example whether it will support superconductivity or not (which is an ontological rather than epistemological issue: i.e. it is a matter of fact about the crystal, whether we know the answer or not). To answer that, you have to describe the details of the crystal structure itself, which is a Level 5 variable. Properties of electrons and ions *per se* simply does not determine anything about such macro level properties (superconductivity emerges at Level 5, and cannot even be described at Level 1). To say "the crystal is made of electrons and ions" does not in principle determine whether such properties will hold or not, in an ontological sense. It is an incomplete characterisation of the system. The issue is:

> **Higher level irreducible variables**: Does the outcome depend on the details of the higher level structure (as in the brain, where the detailed cortical connections matter, or in crystals, where the detailed lattice structure determines whether it is a superconductor) or not (as in a gas, where the details of what molecule is where is immaterial).

An example that makes the point is that a protein may have the chemical formula $H_{n1}C_{n2}O_{n3}S_{n4}N_{n5}$ which (for suitable numbers n_I) lists all its chemical components, but that tells you nothing about its primary, secondary, tertiary, or quaternary structures (Petsko and Ringe 2009), which are all higher level variables. It therefore says nothing about its function.

Given this understanding, the claim I make is that that strong emergence does indeed occur, as discussed below.

6.2.4   Types of Strong Emergence

It is useful to distinguish three types of strong emergence.

- *Unitary emergence*: This name is a generalization of the use of that term in quantum physics. The basic idea is that outcomes at level L_I are determined uniquely by variables at level L_I in terms of initial data given at that level alone. For example, the motion of an ideal pendulum is determined by its initial position and angular velocity. These are macro level variables (Level 6) which entrain billions of lower level variables (motions of particles, Level 2) in a downward way. Again setting the right conditions for superconductivity to emerge (a laboratory level exercise) guarantees that superconductivity will indeed occur (an outcome at both the laboratory and electron levels). Outcomes are reliably determined by initial data.
- *Branching emergence*: This is the case where outcomes at a level L_I are not determined uniquely by initial data of variables at that level because branching dynamics takes place (Ellis and Kopel 2018) with the specific branching that occurs being determined in a contextual way (Noble 2008). For example, the rea-

ding of genes at the cellular level is determined by epigenetic processes influenced by higher level variables (Noble 2011; Kandel 2001). Full knowledge of variables at the molecular biology level (Level 6) at a time t_0 does not determine outcomes at a time $t_1 > t_0$ because of this contextual dependence.

- *Logical emergence*: When intelligent life emerges, rational thought occurs and enables deductive causation (Ellis and Kopel 2018, §6) which has causal powers (Ellis 2005), affecting physical outcomes in a downward way, as do computer algorithms (MacCormick 2011). This is a case of an entirely new kind of causal effect (the causal efficacy of abstract entities) emerging via the structure of the brain and the higher level dynamics it enables.

6.2.5 But does Strong Emergence Happen?

Strong emergence occurs in physics (unitary), digital computers (branching and logical), biology (branching), and the brain (branching and logical), where I refer to the classification in Sect. 6.2.4.

Physics
Strong emergence takes place in both classical and quantum physics. It occurs through *broken symmetry* effects (Anderson 1972) such as chirality and existence of crystal structures (Phillips 2012), leading to quasiparticles (Guay and Sartenaer 2018; Venema et al. 2016) which underlie emergent properties of materials (Tokura et al. 2017) and superconductivity (Laughlin 1999), and *through topological* effects as occurs in the Fractional Quantum Hall Effect (Lancaster and Pexton 2015), topological insulators (Hasan and Kane 2010; Qi and Zhan 2011), and colloids (McLeish et al. 2019; McLeish 2019). Strong emergence in physics is discussed in Sect. 6.4.

Digital Computers
Digital computers provide a definitive example of both branching emergence and logical emergence, through the way computer algorithms control outcomes at all levels right down to the electron level (Ellis and Drossel 2019). This is discussed in Sect. 6.5.

Biology
Branching strong emergence takes place in biology (Ellis and Kopel 2018; McCleish 2017) through downward constraints, downward control, and downward selection enabled by a variety of physiological and molecular biology mechanisms (Noble 2008) involving signaling molecules (Berridge 2014) and the causal efficacy of information (Nurse 2008; Walker et al. 2016), for example that encoded in the gene (Watson et al. 2013). It is also central to the way Darwinian Evolution (in the sense of an extended evolutionary synthesis (Müller 2007; Oyama et al. 2001; West-Eberhard 2003) takes place (Campbell 1974) and enable evolutionary innovation (Wagner 2011), including emergence of new levels in the hierarchy. This is discussed in Sect. 6.6.

The Brain
Agency, based in rational thought, occurs via branching emergence and logical emergence in the brain. This is apparent in the power of thought: the chain of causation from purpose to planning to muscle movement, as discussed in Sect. 6.6.3. This takes place in a social context that enables abstract social constructions such as money and laws to have causal powers (Sect. 6.2.6). Logical emergence takes place (Ellis and Kopel 2018).

6.2.6   Abstract Entities can have Causal Powers

Abstract entities have causal powers. Ideas and plans, social agreements, and Platonic entities all have causal effects in the physical world via brain functioning at the psychological level, enabled by the underlying brain structure and function (Kandel et al. 2013; Frith 2013; Kandel 2012).

Ideas have Causal Powers
Through the functioning of the brain, ideas and plans have causal powers in the physical world. These abstract entities act down to the physical levels to result in billions of atoms being configured as buildings, roads, aircraft, and so on so as to serve individual and social purposes. These outcomes are the result of intention and planning. The ideas and plans that are the key causal element leading to these results are not coarse grained lower level variables. Although they are realised via brain states, they are not the same as any individual's brain state, because they can be shared between people, written down on paper, embodied in computer files, presented in a lecture, and so on. A plan for a building for example is not the same as any particular one of these realisations: it may originate in one person's head, but then attains a life of its own independent of the person who first brought it into being, as it is shared with others and realised in print. Ideas and plans cannot be reduced to any physical variables: they are of a completely different kind because they have a logical and symbolic nature (Deacon 1998).

The Social World
The social world is shaped by many social agreements, which change what happens in social interactions and thus in physical reality.

- *Language* is a symbolic system (Deacon 1998) that is the foundation of social communication. In each society, it is an culturally developed feature that enables society, technology, and commerce to exist. It played a key role in the rise of civilisation (Bronowski 2011).
- *Money* is a key enabler of commercial transactions, as are closed corporations (Harari 2014). It can be realised in many forms, e.g. paper, coins, or electronic.
- *Laws* and associated physical manifestations such as contract documents and passports order society (Harari 2014) and so control physical outcomes, for example passports determine where you can travel.

A very clear example of top-down causation of abstract entities is the rules of games, which constrain what is allowed to happen in physical terms. The rules of chess for example are abstract social agreements that have evolved over centuries; they are not the same as any individual's brain state, although they can be realised in that way, or in verbal form, or in books, or in computer programs. They have causal powers in that they constrain the possible movements of the pieces on a physical chess board, as if they were a force field.

Platonic Entities can have Causal Powers
Mathematical equations (abstract entities) change the world through the human mind, which comprehends them and uses them in engineering design. But what is their nature?

Gödel was a strong supporter of the Platonic nature of mathematics (Wang 1997), as are high level mathematicians such as Roger Penrose (2000) and Alain Connes (1998). The reason is that mathematical facts, such as the distribution of prime numbers, the fact that the square root of 2 is irrational, and Gödel's incompleteness results (Wang 1997), are discovered rather than invented. Competent mathematicians everywhere, whatever their culture, eventually discover them and agree on them. Thus they are universal in nature.

But a long time objection to this proposal has been a claim by some philosophers that if they were to exist, they would be irrelevant as there is no plausible way they could be accessed by the human mind. However Paul Churchland (2013) has given a full answer to this argument, through a study of the way the neural network structure of the brain can learn to recognize abstract patterns. For example the number π—an abstract entity—can be comprehended and calculated to high precision by engineers anywhere in the world, and thereby influence real world engineering outcomes such as the design of aircraft engines and chemical plant. It is not a social construction, it is a mathematical discovery.

6.2.7 Downward Causation Enables Strong Emergence

Strong Emergence Demands Downward Causation
This is clear in the case of biology on the one hand, and digital computers on the other.

Biology
If all the higher emergent levels in the biological hierarchy (Table 6.1) are to each be emergent levels with causal powers in their own right, as claimed by Noble (2011), the effective higher level interactions (such as the pumping of the heart as an integral whole D. Noble 2002) must reach down to lower levels to constrain and shape what happens at those levels (see Fig. 6.1), for instance synthesizing the haemoglobin that is needed for circulation of blood. This happens by cell signaling networks (Berridge 2014) driving the branching logic of metabolic networks and gene regulatory networks at the lower levels according to physiological needs at higher levels, such as the function of the heart (Fink and Noble 2008) or mental

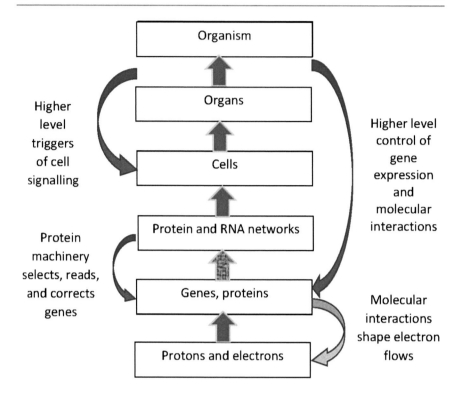

Fig. 6.1 Branching biology causes the underlying physics to branch, via time varying constraints altering the Hamiltonian. (From Ellis and Kopel 2018; extending a figure in Noble 2011)

processes (Kandel 2001). Hence metabolites are produced and proteins synthesized according to higher level needs (Sect. 6.6). But this in turn requires that electron and ion motions at the underlying physical level take place so that these biological functions occur as needed.

Thus the need is that by a process of downwards causation, biology can conscript underlying physics to its purposes. How this happens is discussed in Sect. 6.6.

Computers Conscript Underlying Physics to Implement Abstract Logic

Similarly the flow of electrons in digital computers at the level of transistors is driven by the details of algorithms (Knuth 1973) as expressed in a high level computer programming language. Each level of the emergent hierarchy of virtual machines (Tanenbaum 2006) has a precise effective logic at its own level, as described in the manual for the language (Python, Java, C, Assembly, etc.) used at that level, downwardly controlling electron flows at transistor level according to the digital logic of the corresponding machine code. We can understand every step of this process whereby abstract algorithms are causally effective in the physical world (Sect. 6.5) through existence of application programs[1] that can be contextually driven (Ellis and Drossel 2019).

6.3   How is Downward Causation Possible?

But how is downward causation possible? Many people deny it can happen (see Humphreys 2016; Gibb et al. 2019). In this section, I discuss various ways that downwards causation can take place. It occurs via time independent and time dependent contextual constraints (Sect. 6.3.1), homeostasis/feedback control (Sect. 6.3.2), downward emergence (Sect. 6.3.3), downward adaptation (Sect. 6.3.4), and downward selection (Sect. 6.3.5). I comment on the key features of existence of irreducible higher level variables and higher level organizing principles (Sect. 6.3.6), and the multiple realisation of both at lower levels (Sect. 6.3.7).

6.3.1   Contextual Constraints

All outcomes depend on context. Constraints are a key part of context (Bechtel 2018; Winning and Bechtel 2018) so a mechanism for downward causation is by changing that context (Winning and Bechtel 2018; Blachowicz 2013; Juarrero 1999) Constraints can be physical, electromagnetic, chemical, biological, ecological, or social. A key distinction is whether they are time independent constraints, when outcomes are unitary, or time dependent constraints, when they are not.

Time Independent Constraints
A physical structure such as a dam by its overall construction constrains the position of each particle out of which it is made, but also acts to dam the water. Whether it will in fact successfully dam the water depends on the location of the dam in relation to the topography of the land, and whether it has leaks or not: all high level features relative to the particles making up the dam. They cannot be characterised at a lower level.

An electric circuit is made of a voltage source and wiring that is either open, and no current flows, or closed, and current flows. This is a topological feature of the wiring that cannot be described at any lower level: cutting the loop prevents a current flow. The wiring of a digital computer is an immensely complex construction that links specific transistors, resistors, and capacitors to others in an extremely precise way, determining how electrons can travel between components. The computer will function if and only if this higher level structuring (relative to the level of electrons) is correct.

---

[1] I do not enter here into the debate on the nature of computation (Fresco 2012); the above statement is true whatever view one takes.

Similarly the structure of neural networks in the cortex are a dense connectivity of neurons with each other via axons and dendrites (Kandel et al. 2013). The personality of each individual is determined by the details of that wiring—an immensely complex irreducible topological structure that determines what actions potentials can go where.

In all of these cases, emergence takes place because of the specific constraints arising from those structures. The emergent structures are much more than the particles out of which they are built.

Time Independent Constraints Shape Dynamical Outcomes

The dynamics of a Hamiltonian physical system S with state variables $\mathbf{r}_i$ depends on $\{\mathbf{r}_i, \dot{\mathbf{r}}_i\}$. If it is subject to time independent constraints

$$C(\mathbf{r}_i, \dot{\mathbf{r}}_i) = C_0, \ dC_0/dt = 0 \tag{6.1}$$

the evolution will be unitary: the initial data $\{\mathbf{r}_i(t_0), \dot{\mathbf{r}}_i(t_0)\}$ uniquely determines the state $\mathbf{r}_i(t)$ at all times. A simple example is a frictionless pendulum of length L with $X(t) = L \sin \theta(t)$, $Y(t) = -L \cos \theta(t)$. The constraint is

$$L^2 = X^2 + Y^2 = L_0^2. \tag{6.2}$$

The bob does not fall vertically to the ground because of the constraint (6.2), but rather moves on a circular arc. The equation of motion is

$$\frac{d^2\theta(t)}{dt^2} + \frac{g}{L} \sin \theta(t) = 0, \tag{6.3}$$

and the initial data $(\theta, \dot{\theta})(t_0)$ uniquely determines the outcome $\theta(t)$ at all times.

Time Dependent Constraints

By contrast, if the constraints are time-dependent, it is their time dependence that controls what happens. A dam may have a valve that allows water to flow out when water is needed for agricultural purposes. Outcomes depend on when the valve is open. Similarly an electric circuit will have a switch that determines whether current flows or not, as will a digital computer (its ON/OFF switch); it functions when the switch is on. Brain plasticity at the macro level is based in plasticity at the micro level: details of neural network connectivity and weights in the cortex change all the time in response to interactions with the physical, ecological, and social environment (Kandel 1998; Frith 2013), changing electron flows in the brain and hence mental outcomes.

Time Dependent Constraints Shape Dynamical Outcomes

In the case of Hamiltonian systems, if there are time dependent constraints:

$$C(\mathbf{r}_i, \dot{\mathbf{r}}_i) = C(t), \tag{6.4}$$

it is they rather than initial data that determine outcomes. A simple model is a frictionless pendulum with time varying length $L(t)$. Equation (6.2) is replaced by

$$L^2 = X^2 + Y^2 = L(t)^2, \tag{6.5}$$

and the equation of motion (6.3) becomes

$$\frac{d^2\theta(t)}{dt^2} + 2\frac{\dot{L}(t)}{L(t)}\dot{\theta}(t) + \frac{g}{L(t)}\sin\theta(t) = 0, \tag{6.6}$$

In this case, the initial data $(\theta(t_0), \dot{\theta}(t_0))$ does not determine the solution $\theta(t)$ because of this time-variation of the constraint $L(t)$. It is $L(t)$ that controls the dynamical outcomes because of the second and third terms in (6.6).

This is a very powerful mechanism that underlies why Hamiltonian dynamics is not necessarily unitary dynamics. It occurs in the functioning of digital computers (Sect. 6.5) and in biology (Sect. 6.6); in both cases time dependent constraints conscript the underlying physics to higher level purposes (Fig. 6.1).

Constraints on biological functioning are provided by the environment, for example global climate change (a time dependent large scale effect) has a serious effect on life in the sea (Deweerdt 2017) (an effect on the scale of animals). Global climate change cannot be characterized at any local scale, although its outcomes can.

6.3.2 Downward Control and Homeostasis

A more active process is downward control of lower levels, with feedback control being a key example in biology and engineering.

Downward Control
An engineering example is blasting at a quarry, where a radio transmitter (a holistic macro object) sends signal to a receiver (another emergent macro object) that triggers an ignition device which causes oxygen combination with nitroglycerin at the molecular level. Thus a top down signaling processes to the molecular level causes an explosion that causes rocks to be broken at the macro and micro levels.

In biology, many physiological systems at the macro level (Level 8 in Table 6.1) use cell signaling networks (Level 7) to control metabolic networks and gene regulatory networks shaping molecular biology processes (Level 6), thereby enabling higher emergent levels (Levels 8 and 9) to function according to the logic appropriate to that level (Fig. 6.1). Examples are the heart (Noble 2011) and memory (Kandel 2001).

Feedback Control
A particularly important form of downward control occurs in *feedback control systems*, whereby higher level goals determine lower level outcomes. The system state S is measured by a detector D and compared with a chosen goal G by a comparator C. If $S \neq G$ a signal is sent to a activator A that will alter S so as to move towards

G. The set goal G is the causally effective factor in this situation, determining what happens both at the emergent macro level of the system and the underlying micro-levels as this cycle is continually repeated. The initial data is irrelevant precisely because such a system is designed so as to give the desired output regardless of the initial data (provided the system's parameters are not exceeded). The total system $S = \{S, D, G, C, A\}$ is an emergent irreducible macro entity with a topological configuration physically (it forms a closed loop)

In engineering, this can be implemented in many ways: mechanically, electrically, electronically. The classic example was James Watts' governor for control of the speed of a steam engine. A common example is a thermostat, where the desired temperature T_0 is set on a dial. A thermometer measures the actual temperature T and compares it with T_0. If it is too low, a heater is activated, resulting in the temperature rising (at the macro level) and billions of molecules moving faster (at the micro level)—a classic case of top-down causation. If you set a different temperature on the dial at the macro level, a different outcome results at the micro level. The system as a whole is an irreducible emergent macro system. If you have all the components there but change the topology by undoing one connection, it no longer works. A more complex engineering example is an aircraft automatic landing system.

Homeostasis

In biology, homeostasis is the process of feedback control at all levels whereby an organism can maintain a desired state of equilibrium despite all kinds of disturbances that may occur (Briat et al. 2019; Sauro 2017). It is a fundamental principle of physiology (Randall et al. 2002; Rhoades and Pflanzer 1989), occurring at all levels in the emergent hierarchy: body temperature, blood pressure and so on are maintained at the macro level by using cell signaling networks (Berridge 2014) to control the needed processes at lower biological levels. Homeostasis at those levels, such as cross membrane electrical voltages, levels of potassium ions in axons, and so on operate via small scale localized feedback loops. The networks that implement such feedback in biology are irreducible emergent features (Briat et al. 2019). An example of biological dynamics is a universal biomolecular integral feedback controller for robust perfect adaptation (Aoki et al. 2019).

6.3.3 Downward Emergence

The reductionist paradigm is based on the idea of the existence of lower level entities, such as billiard balls, that have fixed properties independent of the environment. This is often simply not the case when strong emergence occurs. Both the existence of lower level entities (this subsection) and their properties (next subsection) often depend on the environment in key ways. These are important cases of downward causation that strongly contrast to the billiard ball or particle based models.

Quasi-Particles
Properties of metals and semiconductors such as electrical and thermal conductivity
depend on quasi-particles such as phonons (Grundmann 2010). These occur because
collective excitations of the crystal as a whole (Simon 2013), which are by their
nature irreducible emergent entities, lead to existence of quasi-particles at the electron level that are key players in condensed matter physics. Their existence (at the
electron level) is only possible because the discrete symmetry (at the lattice level)
of the crystal structure breaks the continuous symmetry of the underlying physical
interactions (Phillips 2012). Hence they come into being via a process of *downward
emergence* due to the details of the semiconductor material (Grundmann 2010) (they
would not otherwise exist). They are an excellent example of emergence and interactions between levels in physics (Franklin and Knox 2018; Guay and Sartenaer 2018):
the crystal structure is an emergent higher level outcome of ions and electrons, that
causes phonons to come into existence at the lower level (downward emergence)
which underlie dispersion relations and band structure at the macro level (Grundmann 2010), in turn determining optical absorption and electrical resistivity (macro
properties).

Cooper Pairs
A similar example is the Cooper pairs of electrons that make superconductivity possible, which cannot even in principle be explained in a bottom up way (Laughlin
1999). At first glance they should not exists, because electrons repel each other; but
lattice distortions at the crystal level (depending on the nature of the lattice) change
the electric field (at the electron level) and so allow them to come into existence.

Gene Expression
Gene regulatory networks determine the set of proteins that are present in a cell by
controlling what genes will get read when and where. This is a key aspect of developmental biology (Wolpert 2002; Gilbert 2006; Gilbert and Epel 2009), as indicated
in Fig. 6.1. Obviously the physical state at the electron/ion level is changed by gene
regulation processes.

Symbiosis
An important feature of biology is symbiosis. In the case of obligatory symbiosis,
organisms (for example specific birds and flowers exquisitely adapted to each other)
are an emergent irreducible biological entity: its component members cannot exist
on their own. This is for example why the death of bees is a threat to the existence
of many plants.

A major example are multicellular organisms such as human beings. The individual cells in our bodies rely on the body as a whole for their continued existence.
The lungs and circulatory system provides every single cell with oxygen and nutri-
ents that are crucial to its metabolism, and take away waste material. Once the heart
stops beating, blood no longer circulates and all the cells in the body (at the micro
level) die within minutes because they cannot exist on their own; so the macro entity
dies too. The circulatory system is of course itself an irreducible emergent entity:

all its parts must be working and connected in a massively complex topologically connected network in order that the thing as a whole works. Thus for example, for survival, the state of the arteries is as important as the state of the heart.

6.3.4 Downward Modification

Equally important, the nature of lower level entities—how they interact, which characterizes what they are—is often contextually dependent. Downward modification of properties takes place in physics, chemistry, and biology. Downwards selection occurs in engineering and biology, whereby lower level entities are selected to fulfill higher level purposes.

Physics
The behaviour of neutrons is completely different when outside a nucleus than when bound in one (Feldman 2019). Free neutrons decay with a half life of 10 min 11 s, whereas neutrons bound into a nucleus have half lives of billions of years (and if that were not so, we would not be here). Similarly free electrons interact with light in a completely different way than electrons bound in atoms, or in metals.

Chemistry
The behaviour of a sodium atom is completely different when bound into a salt crystal with chlorine, than when free. This is true for all chemical compounds (Hendry 2010).

Development
In developmental biology, cells originate as pluripotent (they can become anything). Their nature gets determined so as to meet specific higher level needs as developmental processes take place. Each cell has its fate determined (whether it becomes a muscle cell, blood cell, neuron, etc.) by positional indicators (morphogens) (Wolpert 2002; Gilbert 2006).

6.3.5 Adaptive Selection

Absolutely crucial to biology is the feature of downward selection, when the nature of lower level entities is determined by a selection process from an initial ensemble $\mathcal{S}$, with the outcome determined by some selection criterion $\mathcal{C}$. The result is a new ensemble $\mathcal{S}^+$ with members that are on average better adapted to the environment, as determined by the selection criterion, than those of $\mathcal{S}$. Thus it is a projection operation

$$\Pi_\mathcal{C} : \mathcal{S} \to \mathcal{S}^+ \tag{6.7}$$

with outcomes dependent on the selection criterion $\mathcal{C}$ (hence it is a top-down effect). The classic physics example is Maxwell's Demon. However it occurs in chemistry and engineering, biology in general and in evolution in particular, and in brain function.

Purification Processes

are key to the possibility of physics and chemistry experiments as well as engineering practice and medicine, because they all demand a supply of pure elements or compounds with specific well-defined properties. Thus there are many separation processes in chemistry, water purification, and in chemical engineering.

Biology

Adaptative selection is a central process in biology. A central feature is variation (Montévil et al. 2016) in order to create an ensemble from which a choice can be made:

> **Environmental adaptation**: It is a profound principle of biology that adaption to the physical, ecological, and social environment takes place at all times and at all levels in a coordinated way. This happens on evolutionary, developmental, and functional timescales. It is a multi-level process whereby communal and individual needs drive adaptation not just of cells and biomolecules but also of developmental systems (cell signaling systems, metabolic networks, and gene regulatory networks) (Oyama et al. 2001; Wagner 2017). This is a process of adaptive selection and hence a specific case of top-down causation (Campbell 1974): different environments lead to different emergent outcomes.

It has functional, developmental, and evolutionary aspects.

Functional Adaptation: Learning

Animals adapt to their environment by learning processes. Brain plasticity at the macro level, entailing learning in response to interactions with the physical, ecological, and social environment, is a key feature of brain function (Frith 2013; Kandel 2012). It is enabled by brain plasticity at the micro level where initially random synaptic connections in the neocortex are adjusted via gene regulation (Kandel 2001): another case of top-down regulation of genes as indicated in Fig. 6.1.

An example of a learning process is a brain that at time t_1 has neural connections encoding knowledge of Maxwell's equations, which were not there at time t_0. This is enabled by a social process of learning: a top-down process from society to detailed cortical connections, which cannot possibly have been determined in a bottom up way—the genome has no knowledge of Maxwell's equations (Ellis and Solms 2017).

Developmental Adaptation: Gene Regulation

Developmental processes in biology take place in a environmentally dependent way (Gilbert and Epel 2009), mediated by developmental systems (Oyama et al. 2001) and *gene regulation processes* which are the explicit mechanisms whereby downward causation takes place to the genome level as indicated in Fig. 6.1. This enables for example acclimatization, whereby individual organisms adjust to changes in the environment.

Evolutionary Adaptation: The Extended Evolutionary Synthesis

Evolution takes place over geological timescales, with adaptive selection taking place repeatedly after replication with variation (Campbell and Reece 2005; Mayr 2002). It occurs as described by the Extended Evolutionary Synthesis (Carroll 2005; Pigliucci and Müller 2000), where evolutionary and developmental processes interact to shape outcomes ("EVO-DEVO" Carroll 2005), leading to emergence of and being shaped by physiology, epigenetics, and developmental systems as well as the genotype, with the developmental systems themselves being shaped by evolution (Peter and Davidson 2011). For example, the existence of vision gives a great adaptive advantage. Evolution consequently lead to development of visual systems at the physiological level that require molecules such as rhodopsin at the molecular level, which would not exist apart from the macro level need of vision in a specific context (for example, eyes for use under water are different from those for use in air Godfrey-Smith 2016).

It should be noted firstly, this is not a gene-centred process, it is much more than that, see *The Music of Life* (Noble 2008) and Evo-Devo writings (Carroll 2005). Secondly, it has major stochastic aspects (Noble 2017), enabling organisms to adapt according to higher level needs (Noble and Noble 2018), so its results are simply not predictable from specific initial conditions. This is also true because cosmic rays, determined by fundamentally random quantum processes, have influenced evolutionary history (Percival 1991). Thirdly, all this means it is highly misleading to describe evolution as an algorithmic process (Dennett 1996, pp. 50 and 63). It is nothing of the sort.

6.3.6   Irreducible Higher Level Variables and Organizing Principles

It is important that downward effects are driven by higher level variables that are irreducible, and are associated with higher level organizing principles that are also irreducible. Higher organizing principles are global states that cannot even in principle be described at any lower level. They reach down to shape what happens at all lower levels.

Physics

Higher level variables in physics that are not reducible to lower level variables are related on the one hand to broken symmetries, such as chirality (Anderson 1972), and on the other to topological features such as occur in polymers and topological insulators (McLeish et al. 2019). These non-local states are discussed in Sect. 6.4.

Autocatalytic Cycles and Sets

According to Hordijk (2013), "[a]n Autocatalytic set is a collection of molecules and the chemical reactions between them, such that the set as a whole forms a functionally closed and self-sustaining system." An example is Bladderwort feeding (Ulanowicz 1995).

A Hiccup

A hiccup is an involuntary spasm of the diaphragm that may occur once off or in a rhythmic series. It is a higher level integral process that can in severe cases have serious consequences such as fatigue and weight loss—clearly downwardly affecting physiological systems at both the macro and micro level.

Biological Organisation

Biological organisation (Mossio et al. 2016) is very complex (Randall et al. 2002; Rhoades and Pflanzer 1989). However there are some fundamental underlying principles such as as closure of constraints (Montévil and Mossio 2015) and of organisation (Mossio and Moreno 2020) which cannot possibly be described or determined at any lower level than the organism as a whole. They are fundamental to the existence and functioning of life. Cellular organisation cannot be described at any lower level (Hofmeyr 2017, 2018).

Being Alive

"Alive" and "dead" are irreducible higher level variables, a state that organises all that happens in a biological system at each moment. This fundamental feature underlies the possibility of Darwinian evolutionary processes (understood in terms of an Extended Evolutionary Synthesis; cf. Carroll 2005; Müller 2007) that lead to the existence of life (Mayr 2002; Campbell and Reece 2005).

Being Conscious

Consciousness is a global brain state, quite different than being asleep. This difference reaches down to affect all aspects of brain and body function (Nichols et al. 2017); the associated Circadian rhythms are a key feature of life (Bass and Lazar 2016). Plans, ideas, and social constructions are irreducible higher level variables associated with consciousness (Sect. 6.2.6).

6.3.7 Multiple Realisation

In real biology, higher-level functions, structure, and variables can be realised in multiple ways at lower levels (Rosen 1958). This multiple realisability (Bickle 2019) causes major problems for any attempt to account for the higher level outcomes in terms of any lower level dynamics, because they cannot be naturally described at those levels (Anthony 2008).

Let higher level $L2$ variables V_I be realizable at lower level L_1 by any one of the combination of lower level variables u_i: $V_I = \cup u_i$ A behavioural law that can be simply stated in terms of the variables V_I at the higher level, such as

"IF {the sun is shining} THEN {the flower will open}"

can only be stated as a series of "OR" statements at the lower level:

"IF $\{u_1$ OR u_2 OR $\cdots$ $\}$ THEN $\{u_{1102}$ OR u_{1022} OR $\cdots$ $\}$"

for a vast number of combinations; one cannot even write them down at the molecular level, where they will number many billions. The latter statement is not a sensible scientific law (basically, it is not expressed in terms of 'natural kinds').

Whenever such multiple realization occurs, this is an indication that downward causation is occurring (Batterman 2018): varying a macro variable causes selection of any one of the equivalence class of lower-level variations that correspond to this higher-level change. A key case is that a vast numbers of different genotypes can produce the same phenotype (Wagner 2017). Darwinian selection takes place in terms of phenotype properties, which then chain down to select any one of the billions of genotypes that result in a better adapted phenotype. Consequently predictable convergence in function has unpredictable molecular underpinnings (Natarajan et al. 2016). In the case of neural networks (Bishop 1995), there are many different detailed connectivity patterns that can result in the same higher-level outcome, such as face recognition. Training the neural network produces any one of those lower level networks that gives the desired macro level performance.

The true causal elements at lower levels are *equivalence classes* that all correspond to the same higher-level elements; these are the 'natural kinds' in terms of which relationships between elements of a field can be defined (Bickle 2019). The bottom line is that causation really happens in terms of the emergent dynamics at the higher level, such as Darwinian evolution (as just discussed in Sect. 6.3.5), whose dynamics cannot be described in lower level terms than survival of individuals. Equivalence classes at the biochemical and physical levels enable it to happen.

6.4   Physics Examples

Although the focus of this essay is on life in general and the mind/brain in particular, it is useful to note that unitary strong emergence and associated top-down causation clearly takes place in the case of condensed matter physics (Drossel 2019) and polymers (McLeish et al. 2019)

Physics is based in unitary Hamiltonian dynamics at the fundamental level, with outcomes determined by the relevant interactions and the initial conditions, together with any constraints that may apply. Unitary emergence takes place through broken symmetries (Sect. 6.4.1) and topological effects (Sect. 6.4.2). In both cases irreducible higher level variables determine physical outcomes. A further key physics case is the emergence of the arrow of time, which cannot be determined in a bottom up way (Sect. 6.4.3).

6.4.1   Broken Symmetries

The equations of fundamental physics are invariant under symmetries which are broken in real-world situations, which is why many emergent properties cannot be deduced in a bottom-up way from the foundational nature of the underlying physics (Anderson 1972). This is a dominant feature of condensed matter physics (Phillips 2012; Simon 2013) and chemistry (Luisi 2002).

Chirality
An emergent feature in physics and chemistry is chirality, that is, the handedness of an entity (Anderson 1972) such as the spin of a particle, the polarization of a wave, or the handedness of the structure of a molecule. This has important outcomes in biology, where chirality affects biological activity because naturally occurring amino acids and sugars are chiral molecules. Thalidomide is a key case in point: the left-handed molecule was fine, but the right-handed one caused major abnormalities in babies. Chirality is an emergent property that cannot be determined locally: it needs some comparison reference object in order to be determined.

Quasiparticles
Physics of quantum materials (Keimer and Moore 2017) is based in the way the continuous symmetry of the underlying fundamental theory is broken by the discrete crystal symmetry (Simon 2013), leading to quasiparticles that control electrical and thermal conductivity and optical properties. This is a case of downward emergence (Sect. 6.3.3).

6.4.2   Topological Effects

Physical systems characterised by topological ordering are strongly emergent because the relevant variables are non-local variables whose values are not determined by any local properties (McLeish et al. 2019). They occur at both micro and macro levels.

Quantum Examples
The *fractional quantum Hall effect* (FQHE) is an example of topological emergence (Lancaster and Pexton 2015; McLeish et al. 2019), where fractionally charged particles arise out of collective behaviour resulting from magnetic field interactions with 2-dimensional systems of electrons. Topological insulators—a vibrant field of current research—are another fascinating example of strong emergence in quantum physics with downwards effects (Hasan and Kane 2010; Qi and Zhan 2011),

Polymers and Colloids
Soft matter physics (McLeish et al. 2019; McLeish 2019) deals *inter alia* with polymers and colloids, characterised by topological variables at a higher level than the electron level (Lancaster and Pexton 2015).

Knots and Knitting
Knitting is an extraordinary operation where a 1-dimensional polymer chain is built into a 2-dimensional fabric (Fig. 6.2) and then into a three-dimensional garment. Multiple levels of topological entanglement are thereby created: at the polymer level,

Fig. 6.2 Detail of structure of knitted fabric at stitch level (left) and outcome at fabric level (right). Irreducible higher level variables are introduced at each level, because they are topological (Wikipedia)

at the fibre level, at the stitch level, at the fabric level, and at the garment level. The same is essentially true for knots in ropes, as used in sailing, climbing, and so on. They cannot emerge bottom-up: they are all the result of intelligent thought.

6.4.3 Direction of Time

The direction of time is a key property of macro physics, chemistry, and biology. However given all the details of positions and momenta of particles in a cylinder, you can't tell from that data in which direction of time entropy will increase, because coarse-graining a time symmetric micro theory in an isolated system necessarily results in time-symmetric macro physics. Whatever bottom-up proof you have via coarse graining that entropy S increases with time t: $dS/dt \geq 0$, will also prove that entropy increases in the opposite direction of time $t' := -t$ because the identical proof will show $DS/dt' \geq 0$. This is true both for the classical proof (Boltzmann's H-theorem) and the quantum field theory version (Weinberg 1995, pp. 150–151).

A global condition (the "Past Condition", Albert 2003) is required to set initial conditions for local arrows of time, indicating *inter alia* in which direction of time entropy will necessarily increase. This is related to the cosmological *Direction of Time* (Ellis 2014; Ellis and Drossel 2020)—a global variable deriving from the evolution of the universe, which can't be determined from lower level variables as it relates to the evolution of the Universe as a whole, and so is strongly emergent. *Inter alia* it determines the quantum mechanical arrow of time (Drossel and Ellis 2018).

6.5 Digital Computers

Digital computers are an excellent example of branching emergence and associated downward causation, because we can understand everything that goes on in them (since we built them!) Their structure enables the causal power of algorithms (Sect. 6.5.1). The link to the underlying physics occurs through logical branching via transistors (Sect. 6.5.2)

6.5.1 Causal Power of Algorithms

Digital computers are driven by the abstract logic of algorithms (Knuth 1973), initially coded in a specific high level language (Abelson and Sussman 1990), and then chained down from the top level of the software hierarchy (the tower of virtual machines Tanenbaum 2006) to machine code level by compilers (Aho et al. 2006) or interpreters. They are realised in a different language at each level, and at the machine language level control the hardware (transistors are "ON" or "OFF") through binary code ("0" or "1"). Because on the logical side arbitrary computations can be expressed this way (Turing's great discovery Hodges 1992), and on the hardware side the basic Boolean operations ("AND", "OR", "NOT") can be realised by suitable combinations of transistors,[2] this results in the ability to carry out any computable task. Hence algorithms can and do change the world (MacCormick 2011).

In digital computers, the electrons in the transistors flow in accord with the logic of a chosen algorithm (translated into binary code). Physics does not determine those algorithms, precisely because they are logical in nature. They are abstract entities (logical procedures) that are causally effective.

That this is a process of downward causation is obvious (the computer program is loaded by the operator at the macro level; different algorithms result in different electron flows at the transistor level). This downward causation is reflected in the multiple realisability of what happens at each downward step in the tower of virtual machines (Tanenbaum 2006); for example the Java Virtual Machine (Lindholm et al. 2014) enables Java to run on any hardware, and even the hardware/software distinction is mutable (Tanenbaum 2006). This enables the branching emergence whereby higher levels (such as a Word Processor program run at the top level) have real causal powers and control what happens at every physical level in the computer, including the electron level.

6.5.2 Logical Branching via Transistors

But how is this downward causation possible, leading to branching causation, given the alleged unitary nature of the underlying physics? On the one hand, algorithms control what gates are operated in what sequence hence the electrons at the bottom level do what they are told to do by the algorithms, all the while obeying Maxwell's and Newton's equations. On the other hand, the description on the microscopic level is based on a Hamiltonian for the ions and electrons (Phillips 2012, p. 16):

---

[2]This is required in order that a Universal Turing Machine can function.

$$H(\mathbf{R}_i, \mathbf{r}_i) = -\sum_i \frac{\hbar^2}{2M_i} \nabla^2_{\mathbf{R}_i} - \sum_i \frac{\hbar^2}{2m_e} \nabla^2_{\mathbf{r}_i} + \sum_i \sum_{j>i} \frac{Z_i Z_j e^2}{4\pi \epsilon_0 |\mathbf{R}_i - \mathbf{R}_j|}$$
$$- \sum_i \sum_j \frac{Z_i e^2}{4\pi \epsilon_0 |\mathbf{R}_i - \mathbf{r}_j|} + \sum_i \sum_{j>i} \frac{e^2}{4\pi \epsilon_0 |\mathbf{r}_i - \mathbf{r}_j|}, \qquad (6.8)$$

where $\mathbf{R}_i$ are the positions of the ions with atomic weight Z_i and $\mathbf{r}_i$ the positions of the electrons. By itself, this would indeed lead to unitary dynamics.

However we must add to Hamiltonian a time dependent term $V(\mathbf{r}_i, t)$ due to the applied gate voltage $V(t)$ that turns the transistor ON or OFF (Ellis and Drossel 2019). This leads to a potential energy term in the Hamiltonian of the electrons:

$$H_V(t) = \sum_i eV(\mathbf{r}_i, t) \qquad (6.9)$$

where the Level 5 (see the left-hand column of Table 6.1) variable $V(t)$ determines the Level 2 variables $V(\mathbf{r}_i, t)$ in a downward way. This leads to a displacement of the electrons until a new equilibrium is reached where the electrical field created by the modified charge distribution cancels the electrical field due to the gate voltage. In order to calculate this new equilibrium, a self-consistent calculation based on the charge density due to doping, gate potential, and thermal excitation must be performed.

Thus if we regard the electrical field as a time-dependent constraint on the electrons in the transistor, this is a form of what is discussed in Sect. 6.3.1. The causal power of algorithms is realised via details of transistor design whereby the change in $V(t)$ enables currents to flow or not in the transistor, so it can act as part of a logical gate enabling higher level logic to emerge from the digital logic at the transistor level (Mellisinos 1990; Ellis and Drossel 2019).

6.6   Biological Examples

Biology emergence is based in contextual branching at each level, for example in gene regulatory networks and metabolic networks at the lower levels (Sect. 6.6.1). This enables emergence of physiological structures such as the heart (Sect. 6.6.2) and brain (Sect. 6.6.3). The link to the underlying physics is the contextual effects enabled via biomolecules, which enable top-down control of the underlying physics (Sect. 6.6.4).

6.6.1 Biology and Contextual Branching

The structure and function of biology are closely intertwined (Campbell and Reece 2005; Randall et al. 2002; Rhoades and Pflanzer 1989).

Structure
The biological structural hierarchy is shown on the right in Table 6.1. The cell is the crucial level: all living systems are made of cells, which are pluripotent to begin with but then (in multicellular animals) are specialized to serve specific functions by developmental processes (Gilbert 2006; Wolpert 2002; Gilbert and Epel 2009) (Sect. 6.3.3).

Function
Causation at each level of the biological hierarchy tends to further the function α of a trait T through contextually informed branching dynamics so as to enhance the overall viability of the organism in its environment. As summarised by Hartwell et al.:

> Although living systems obey the laws of physics and chemistry, the notion of function or purpose differentiates biology from other natural sciences. Organisms exist to reproduce, whereas, outside religious belief, rocks and stars have no purpose. Selection for function has produced the living cell, with a unique set of properties that distinguish it from inanimate systems of interacting molecules. Cells exist far from thermal equilibrium by harvesting energy from their environment. They are composed of thousands of different types of molecule. They contain information for their survival and reproduction, in the form of their DNA. (Hartwell 1999)

Because both the external environment and the internal milieu are continually changing, adaptation must take place on an ongoing basis (Sect. 6.3.5). Consequently, *Contextual Branching Dynamics* (Ellis and Kopel 2018) is required to attain desired outcomes. A variety of systems enable this to happen, at both the macro and micro levels.

Systems
At the macro level, all the physiological systems required for bodily function (structure) are controlled so as to respond appropriately to environmental circumstances (function). These systems include the Circulatory System, Immune System, Nervous System, Sensory Systems, and so on (Randall et al. 2002; Rhoades and Pflanzer 1989). They are emergent systems with the ability to respond appropriately at their emergent level according to their function, because that is what they are structured to do, thanks to evolutionary and developmental processes. They are supported at the molecular biology and cellular levels by

- *Metabolic Networks*, controlling production, distribution and use of matter and energy, and disposal of waste products (Wagner 2017)

- *Gene Regulatory Networks* controlling the reading of the genotype so as to produce proteins needed to construct the phenotype (Wagner 2017)
- *Cell signaling networks* conveying information that controls the gene regulatory networks and metabolic networks (Berridge 2014)
- *Developmental systems* that coordinate developmental programs generating an adult organism from a single cell (Gilbert 2006; Wolpert 2002)

These often proceed on the basis of the lock and key molecular recognition mechanism of supramolecular chemistry (Lehn 1993, 1995), with branching dynamics controlled by transcription factors, enzymes, and so on. In this case there is a selective response to a particular signaling molecule (Berridge 2014). However they may also respond to physical stimuli, as for example in the case of voltage gated ion channels.

As a particular example, transcription factors may be ON (that is, able to bind to DNA) or OFF, in this way controlling transcription of DNA to messenger RNA and so to proteins needed for cell function. Thus if transcription factor TF_2 modulate synthesis of proteins in a metabolic pathway, it embodies branching logic of the form

$$IF \; \{TF_2 \; on\}, \; THEN \; \{X_2 \rightarrow X_3\}, \; ELSE \; NOT \qquad (6.10)$$

where X_A are metabolites (Goelzer et al. 2008). This is what allows contextual control of gene expression (Noble 2011) (Sect. 6.2.2): regulatory processes determine what gene gets turned on where and when. This contextuality of branching represents top-down effects (Noble 2011; Kandel 1998).

Such processes are hierarchical and modular (Ravasz et al. 2002; Goelzer et al. 2008): a higher level regulator TF_1, sensitive to macro variables such as blood pressure or heart rate, can modulate the synthesis of intermediate enzymes and local transcription factors (such as TF_2 in Eq. 6.10), enabling top-down control of the process. The dynamics of such modules is multiply realizable: it does not matter what the internal variables and dynamics is, as long as the resultant genes or metabolites are what are required. Branching emergence occurs.

These processes all of course must obey thermodynamic constraints. However significant as they are (Jeffery et al. 2019), by themselves thermodynamic or entropic considerations are far from characterizing life, as they do not entail the idea of function or purpose commented on by Hartwell et al. (Hartwell 1999), nor complexities such as closure of constraints (Montévil and Mossio 2015).

6.6.2   Physiology: The Heart

The heart is a crucial macro system of the body (Sect. 6.3.3) which is an irreducible whole (Sect. 6.3.6), for example a key role in its functioning is played by a cardiac pacemaker which is a complexly connected set of cells controlling the heart rhythm.

The heart has been modeled in depth by Denis Noble (2002), who calculated transmural pressure acting on coronary vessels due to myocardial stress: a downward effect from non-local variables to the level of coronary vessels, which respond to those stresses. His studies show how there is contextual regulatory control of lower level biological processes by higher level physiological states (Noble 2011). Noble et al. (2019) refer to the following data on how lifestyle choices influences RNAs and so control gene expressions:

- Bathgate et al. (2018) have shown how "RNA levels of control are changed by the lifestyle choices" in identical twin studies.
- D'Souza et al. (2017) investigated the fact that "athletes have lower heart rates than non-athletes, which was once attributed to greater vagal tone. The changes have now been traced to microRNAs that downregulate expression of the HCN gene, so that the depolarizing current (if) produced in the sinus node cells is reduced by as much as 50%".

Thus these studies work out in detail how high-level choices produce change at the molecular level, thus demonstrating the downward causal power of higher level choices.

6.6.3 Structure and Function of the Brain

The power of thought (intentions, plans, equations) was discussed in Sect. 6.2.6. The issue is how this can emerge from the underlying physics. The context is the Central Nervous System hierarchical structure (Kandel et al. 2013). This is the physical basis of consciousness. The neuron is the cellular level. The neocortex has cortical columns characterised by layered dense interconnections of neurons joined by synapses. The differences between people lie in the details of this neural network structure, which is determined by adaptation to the environment (neural plasticity underlies memory and learning). Action potentials propagate down neuron dendrites to the soma (nucleus) and then down axons to synapses where connection is made with other neurons by neurotransmitter diffusion across the synaptic cleft. Action potential spike chains are enabled by flow of ions in and out of axons via voltage gated ion channels, to create a current flow along the axon. This gives the Hodgkin-Huxley equations.

When built into neural networks linked by synapses in the neo-cortex, the resulting action-potential spike chains are the basis of logical thought and other mental phenomena. We do not know how thoughts are coded in action potential spike chains, nor do we know how consciousness arises; possibly by non-local synchronization of neurons (Kanter et al. 2011). However, Eric Kandel (1998) gives a clear set of principles underlying what happens, as follows:

1. All mental processes derive from operations of the brain.
2. Genes determine neuronal functioning.
3. Social and developmental factors contribute importantly to the variance in mental illness. These factors express themselves in altered gene expression.
4. Nurture is ultimately expressed as nature.

5. Altered gene expression induced by learning gives rise to changed patterns of neuronal connections, which give rise to different forms of thinking and behaviour.
6. Psychotherapy produces changes in long-term behaviour by learning which produces changes in gene expression, and hence changes in neuronal interconnection.

These principles express top down action from the mental level to details of neural connections via gene regulatory networks (as in Fig. 6.1). Brain plasticity at the macro level is enabled by changes in synaptic weights at the micro level, based in experience, via suitable gene expression. Logical emergence (Gödel's concern; Wang 1997) occurs (Ellis and Kopel 2018).

6.6.4   The Link to Physics: Contextual Effects via Biomolecules

The branching logical function that emerges in the brain is enabled at the molecular level by particular proteins, and again (as in Sect. 6.5) the issue is how is this branching dynamics compatible with the allegedly unitary underlying dynamics. I consider first *Ligand gated ion channels*, and then *Voltage gated ion channels*. Both control flow of ions across membranes, hence facilitating messaging in the neural system (Magleby 2017).

Ligand Gated Ion Channels
These are key to synaptic function (Unwin 1993), due to their specific molecular structure (Fig. 6.3, left). They occur on the postsynaptic neuron in a synaptic cleft. Neurotransmitters released into the cleft by an excited presynaptic neuron binds to them if of the specific type they recognize, causing a conformational change which opens the ion channel. The resulting flow of ions across the cell membrane leads to either depolarization or hyperpolarization and so controls spike chain initiation. Thus time dependent molecular signals (Berridge 2014) reach down (Noble 2011; Kandel 2001) to change the conformation of biomolecules and alter outcomes. Recognition

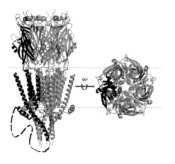
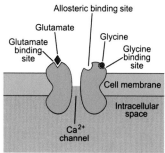

Fig. 6.3 *Ligand gated ion channels.* Left: Basic structure of the cation-selective pentameric ligand-gated ion channels (Wu et al. 2015). Right: Stylized depiction of an activated N-methyl-D-aspartate (NMDA) receptor (Wikipedia)

of the specific ligand by the receptor is due to the lock and key molecular recognition mechanism of supramolecular chemistry (Lehn 1993).

In one specific case (Fig. 6.3, right), the ligand-gated ion channel is gated by the simultaneous binding of glutamate GLU and glycine GLY, thus it acts as an AND gate. The ion channel structure results in branching dynamics with the following logical structure:

$$\text{IF } \{\text{GLU AND GLY}\} \text{ THEN } \{\text{allow ion flow}\}, \text{ ELSE not} \quad (6.11)$$

This logical function is enabled by changes in the 3-dimensional conformation of the ion channels (Fig. 6.3). This is the way the underlying physics (which determines possible molecule shapes) enables the logical (binary) outcome expressed in (6.11).

Thus control is via conformational change of proteins, which changes dynamics (Grant et al. 2010). But how does this relate to the underlying Hamiltonian dynamics? The ligand binding changes molecular shape and hence alters the Hamiltonian and hence the dynamics, as in the case of the pendulum with varying length (Sect. 6.3.1). Martin Karplus states:

> First, evolution determines the protein structure, which in many cases, though not all, is made up of relatively rigid units that are connected by hinges. They allow the units to move with respect to one another. Second, there is a signal, usually the binding of a ligand, that changes the equilibrium between two structures with the rigid units in different positions. (Karplus 2014)

Working this out needs detailed quantum chemistry simulations with many-body Hamiltonians for electrons and nuclei. The Schrödinger equation for the ions and electrons comprising the biomolecules is again (6.8). Solving it in detail for a very large number of nuclei can be done by the CHARMM simulation suite of programs used widely for macromolecular mechanics and dynamics (Brooks et al. 1983, 2009). However one can get a heuristic solution using the Born-Oppenheimer approximation (King et al. 2010) to determine the dependence of outcomes on nuclei distances $\rho_{IJ}(t)$, which can be regarded as time-dependent constraints dependent on binding molecules. This is how messenger molecules alter the outcomes of the underlying physics (which electrons flow where and when).

Voltage Gated Ion Channels

These ion channels (Ranjan et al. 2019; Catterall 1995) are crucial to spike train propagation. When imbedded in axon and dendrite membranes they control the flow of potassium, sodium, and chloride ions across the membrane, leading to action potential spike chain propagation along the axons and dendrites. They implement the following branching logic:

$$\text{IF } \{V > V_0\} \text{ THEN } \{\text{allow ion flow}\}, \text{ ELSE not} \quad (6.12)$$

via conformational changes of these molecules induced by the membrane potential V, for some threshold V_0 (Ellis and Kopel 2018). These proteins are selected in order to perform this function via Darwinian adaptive processes (Wagner 2017). When built into neural networks with neurons linked by synapses in the neo-cortex, the resulting spike chains are the basis of logical thought and other mental phenomena (Kandel et al. 2013; Kandel 2012; Frith 2013). Thus in this case, branching emergence supports logical emergence (Sect. 6.2.4), as illuminatingly discussed by Paul Churchland (2013).

In both cases,
conformational change of ion channels enables lower level branching dynamics (Ellis and Kopel 2018): biology causes physical branching by altering constraints at the molecular level in a time dependent way (Sect. 6.3.1). In this way unitary physics is conscripted to implement the branching logical dynamics of biology such as in Eqs. 6.11 and 6.12, enabling branching emergence (Sect. 6.2.4) to occur.

6.7   Branching Physics and Supervenience

The previous sections give sound arguments firstly that strong emergence does indeed happen, and secondly regarding how it happens. However arguments based on supervenience together with the alleged causal completeness of physics at the lower levels claim this is not possible (Kim 1998, 1999), as discussed in depth in Gibb et al. (2019). What answer can one give?

The premise is wrong. Physics is not causally complete firstly, because of quantum uncertainty (Sect. 6.7.1). Secondly, there are no isolated systems in the real world (Sect. 6.7.2). Thirdly, physics by itself is not causally complete because of the contextual effects discussed in previous sections (Sect. 6.7.3). Consequently while synchronic supervenience may be true, in most real world situations diachronic supervenience is not (Sect. 6.7.4); therefore arguments from supervenience fail to disprove strong emergence. The conclusion considers the implications of all the above for emergence and reductionism (Sect. 6.8).

6.7.1   Physics is not Causally Complete: Quantum Uncertainty

Physics is not causally complete at lower levels because it is not possible *in principle* to predict specific outcomes at the quantum level (Lucas 1996; Ghirardi 2007). This is proven *inter alia* by the foundational 2-slit experiment (Fig. 6.4). Quantum uncertainty is irreducible in the one real world where we can carry out experiments, so there is a fundamental indeterminism of the physical universe (Santo 2019). Statistical outcomes however are determinate.

The point then is that quantum uncertainty can get amplified to the macro level, and this happens for example in biology in the case of radiation damage to DNA caused by cosmic rays (Percival 1991); but the emission of a cosmic ray by an excited atom is a quantum event that is intrinsically unpredictable (there is no physics equation that tells when it will be emitted, or in what direction it will go). Cosmic ray damage

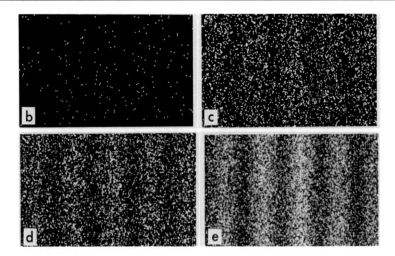

Fig. 6.4 *Quantum uncertainty.* Double slit experiment performed by Tonomura showing the build up of an interference pattern of single electrons. The numbers of electrons are, **b** 200, **c** 6000, **d** 40,000, and **e** 140,000

to DNA has arguably been a significant effect in terms of the evolutionary history of life on Earth (Scalo et al. 2001).

Amplification of quantum events also takes place in photomultipliers, CCDs, particle detectors, and so on; indeed one can relate this to Eq. 6.6 for a variable length pendulum as follows: consider a Schrödinger-cat like setup, where a radioactive element emits particles received by a detector which each time sends a signal to a computer that uses it to alternately increase and decrease the length $L(t)$. Then the dynamical outcome of Eq. 6.6 is in principle unpredictable: the motion of the pendulum is not determinate.

Thus the claim of causal completeness of physics, in the sense of being a unitary theory where specific outcomes are predicted by the initial data, is simply not correct.[3]

6.7.2   There are no Isolated Systems

The belief that physics leads to unitary dynamics is based on the combination of Hamiltonian dynamics with the concept of an isolated system. But while the isolated systems that would lead to unitary behaviour at the micro and hence macro level are a useful conceptual device to isolate causal mechanisms at work, they do not in fact exist in the real universe, both in temporal and causal terms. They may however be a useful approximation in restricted circumstances for a limited timespan.

---

[3]I am discounting *Many Worlds* and *Hidden Variable* theories because they simply have no cash value for the physicist doing experiments in her laboratory. They do not predict, on the basis of the initial data, the outcomes she will measure in specific individual cases.

A first example is that a freely oscillating pendulum in a laboratory cannot have existed for all time: it will not have existed before it was manufactured. Furthermore, however excellent it is, it will in fact not be frictionless: it will gradually slow down due to friction, which is possible only because the laboratory is in touch with a heat sink (the dark night sky) into which it dissipates waste heat. The reason the night sky is dark is because the cosmological context of the expanding universe is such that Cosmic Background Radiation has a temperature of 2.7 K (Peter and Uzan 2013).

A second example is that a digital computer has a non-zero error rate due to cosmic rays (Ziegler and Lanford 1979). The abstract of O'Gorman et al. states:

> This paper presents a review of experiments performed by IBM to investigate the causes of soft errors in semiconductor memory chips under field test conditions. The effects of alpha-particles and cosmic rays are separated by comparing multiple measurements of the soft-error rate (SER) of samples of memory chips deep underground and at various altitudes above the earth. The results of case studies on four different memory chips show that cosmic rays are an important source of the ionizing radiation that causes soft errors. (O'Gorman et al. 1996)

As indicated in the previous section, the specific resultant errors that occur (at the macro level) due to ionizing effects (at the micro level) are not predictable even in principle.

Random Environmental Effects at the Molecular Level
In practice, the environment for biological systems at molecular levels is highly random: they are subject to massive fluctuations due to random molecular motion. However molecular machines have evolved to give reliable outcomes in this context by harvesting the molecular storm, see *Life's Ratchet: How Molecular Machines Extract Order from Chaos* by Peter Hoffmann (2012). A key way higher level layers can extract order out of this chaos is by adaptive selection from the ensembles provided by random processes, from which they can select preferred outcomes according to higher level selection criteria and thereby harness stochasticity (Noble and Noble 2018) (Sect. 6.3.5). In particular this plays a key role in brain function (Glimcher 2005; Rolls and Deco 2010). Thus far from physics being unitary at the appropriate level, as envisaged in supervenience discussion, it is highly random at this level, and biology takes advantage of this feature.

6.7.3 Physics by Itself is not Causally Complete

In addition, physics by itself is also not causally complete because of the contextual effects that determine outcomes, as discussed in this essay.

No Physical System is Isolated from its Larger Context
Physics *per se* is not causally complete because biological, psychological, social, and environmental processes affect what happens in the world according to higher level dynamics, thereby jointly shaping outcomes. They do this by altering the context

within which specific physical outcomes occur. Physical forces do the work needed in this larger functional context. I will just mention two specific cases: Darwinian evolutionary processes, and the functioning of digital computers.

Darwinian Evolution

Evolution over geological timescales in the sense of an extended evolutionary synthesis (Pigliucci and Müller 2000) is a key process in biology (Mayr 2002) that cannot be comprehended in physics terms (neither 'animal' nor 'alive' are physics concepts). It is a form of adaptive selection (Sect. 6.3.4): higher level conditions select what happens at lower levels according to the selection criterion of successful reproduction. Now the point is that survival of individuals depends on factors like evolution of the social brain (Dunbar 2003; Gintis 2011) and development of the symbolic capability of the human mind (Deacon 1998). These are both irreducible higher level factors that have lead to the extraordinary success of the human race, in particular enabling the emergence of technology and commerce and thereby altered physical outcomes across the world (Bronowski 2011; Harari 2014). The underlying physical interactions in the brain and outcomes in the world only proceed within this broad context of human evolutionary development, in which abstract elements such as national pride and societal issues such as technological competence play an important role. Gene-cuture co-evolution takes place (Gintis 2011); physical causation is just one part of the overall story (with some key causal factors being abstract).

Digital Computer Outcomes

Digital computers depend on the algorithms that drive them (Sect. 6.5). The algorithms deployed depend on goals for which computer programs are written, which are engineering, economic, and social purposes, for example including the development of search engines and social media. These in turn are crucially affected by the values and understanding of meaning that are driving them (Thompson 2019; Ellis and Drossel 2019), which are in fact the highest level variables shaping the outcomes of digital computers.

In Summary

Physics *per se* is not causally complete. In determining what actually happens, all these other factors need to be taken into account as part of the causal matrix determining specific physical outcomes. Physics alone cannot determine them (Ellis 2005).

6.7.4 Synchronic and Diachronic Supervenience

It has been claimed (Kim 1998, 1999) that supervenience prevents strong emergence; arguments for and against are discussed in Gibb et al. (2019).

Unitary Emergence

Supervenience assumes the lower level state $L_1(t)$ synchronically (i.e. at each instant t_0) uniquely determines the higher level state $L_2(t)$: $L_1(t_0) \Rightarrow L_2(t_0)$. If $L_2(t_0)$ were

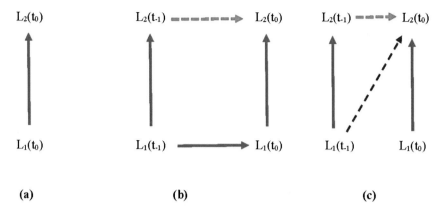

Fig. 6.5 *Unitary supervenience.* **a** Synchronic supervenience. The lower level state $L_1(t_0)$ uniquely determines the higher level state $L_2(t_0)$ at that time. **b** The lower level state $L_1(t_{-1})$ uniquely determines the lower level state $L_1(t_0)$, which uniquely determines the higher level state $L_2(t_0)$ at that time. **c** The outcome: unitary diachronic supervenience. The lower level state $L_1(t_{-1})$ at time t_{-1} uniquely determines the higher level state $L_2(t_0)$ at the later time t_0

different, $L_1(t_0)$ would be different (Fig. 6.5a). In the case of unitary emergence, physics is indeed causally complete at the lower level, so $L_1(t_{-1}) \Rightarrow L_1(t_0)$ (Fig. 6.5b), and then there is no freedom for higher levels *per se* to influence outcomes: unitary supervenience $L_1(t_{-1}) \Rightarrow L_2(t_0)$ occurs (Fig. 6.5c). Both synchronic and diachronic supervenience hold: the lower level state $L_1(t_{-1})$ determines the lower level state $L_1(t_0)$ and hence both the higher level states $L_2(t_{-1})$ and $L_2(t_0)$.

However as discussed in depth above, in living systems we have branching emergence rather than unitary emergence, and in the case of the brain we have logical emergence (Sect. 6.2.3). The outcome then is very different.

Branching Emergence

In the case of non-unitary emergence, physics is not causally complete at the lower level (Sects. 6.7.2 and 6.7.3) and higher levels crucially influence outcomes, so non-unitary diachronic supervenience occurs. Synchronic supervenience means the lower level state $L_1(t_{-1})$ determines the higher level state $L_2(t_{-1})$ (Fig. 6.6a), and $L_1(t_0)$ determines $L_2(t_0)$ (Fig. 6.6b). However the lower level evolution $L_1(t_{-1}) \rightarrow L_1(t_0)$ depends on the higher level state $L_2(t_{-1})$ via time dependent constraints, downward emergence, and downward selection (Sect. 6.3). Thus the higher level state $L_2(t_{-1})$ influences $L_1(t_0)$ and hence $L_2(t_0)$ (Fig. 6.6c): the higher level state has causal power. Diachronic supervenience does not occur (Sartenaer 2015).

Example: Learning

Many examples have been given above, and I will consider just one here: the brain of a student who knows Maxwell's equation at the time t_0. If we could reproduce in another brain in precise detail all her neural connections together with all the excitations of of those neurons, it is plausible that other brain would also know

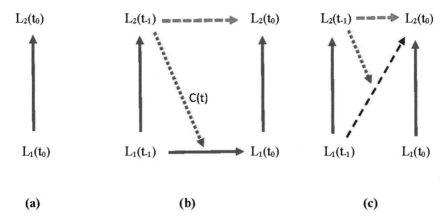

Fig. 6.6 *Non-unitary diachronic supervenience.* **a** Synchronic supervenience. The lower level state $L_1(t_0)$ uniquely determines the higher level state $L_2(t_0)$ at that time. **b** The lower level state $L_1(t_{-1})$ does not uniquely determine the lower level state $L_1(t_0)$, which is also influenced by the higher level state $L_2(t_{-1})$. **c** The outcome: non-unitary diachronic supervenience. The lower level state $L_1(t_{-1})$ at time t_{-1} does not uniquely determine the higher level state $L_2(t_0)$ at the later time t_0. In fact $L_2(t_{-1})$ determines the outcome

Maxwell's equations (Fig. 6.6a). However the brain is plastic: at a previous time t_{-1} she did not know Maxwell's equations; the question is how did they get coded in her synaptic connections at the later time?

This happened by a process of learning at the macro level. Through interactions with a teacher, textbooks, and her own internal thoughts, she learned those equations by time t_0. The learning process proceeded by downward causation controlling the reading of genes in synapses (as stated in Kandel's principles Kandel 1998; Sect. 6.6.3), and so altering her neural network connections: the lower level state $L_1(t_{-1})$ changed to $L_1(t_0)$. That change could not take place without the explicit learning process indicated by the downward arrow $C(t)$ in Fig. 6.6b (see the discussion in Kandel 2001), which is a socially mediated process (no brain can meaningfully be regarded as living in isolation; Donald 2002; Dunbar 2003).

Overall, one should note that in any case "supervenience is an inadequate device for representing relations between different levels of phenomena" (Humphreys 1997). Emergence involves much more.

6.8   Conclusion: Emergence and Reductionism

This essay has made the case that the reductionist views Gödel opposed (Wang 1997) are unjustified. On the contrary, unitary strong emergence, branching strong emergence, and logical strong emergence all occur. Contextual choices are being made all the time at all levels of the emergent hierarchy of biology (Table 6.1), with real causal power residing at every level, including the the psychological and social levels. All levels are equally real, as emphasized by Noble (2011). Concepts exist as much as electrons do; enzymes do not allow for a mechanical interpretation of all functions of the mind (Wang 1997, pp. 174 and 192).

The result is amazing:

> We can give a general characterization of what it is for a system to be able to represent within itself some other system, and so can think of organisms in terms not of biochemistry or evolutionary biology but of information theory and formal logic. And from this point of view we can consider not only consciousness but self-consciousness, and a system that can represent within itself not just some other system but itself as well. There are a whole series of self-reflexive arguments. (Lucas 1996)

Top down action enables this, whereby the lower level physics is conscripted to fulfill higher level purposes via time dependent constraints, downward emergence, and downward selection. Thus the arguments against strong emergence based in diachronic supervenience do not hold. There is no violation of the underlying physical laws. Rather their operating context is shaped to obtain the desired outcomes.

What questions reductionists cannot answer

Reductionists cannot answer why strong emergence (unitary, branching, and logical) is possible, and in particular why abstract entities such as thoughts and social agreements can have causal powers. The reason why they cannot answer these questions is that they do not take into account the prevalence of downward causation in the world, which in fact occurs in physics, biology, the mind, and society.

Details of why this is so have been given above. Further support for this view comes from consideration of both quantum entanglement (Passon 2019), and the emergence of classical physics from quantum theory (Ghirardi 2007; Ellis and Drossel 2019), which I do not consider here.

References

Abelson, H., & Sussman, J. S. (1990). *Structure and interpretation of computer programs.* Cambridge: MIT Press.

Aho, A. V., Lam, M. S., Sethi, R., & Ullman, J. D. (2006). *Compilers, principles, techniques, and tools.* Reading: Addison Wesley.

Albert, D. (2003). *Time and chance.* Cambridge: Harvard University Press.

Alon, U. (2006). *An introduction to systems biology: Design principles of biological circuits.* London: Chapman and Hall/CRC.

Anderson, P. W. (1972). More is different. *Science, 177,* 393–396.

Anthony, L. M. (2008). Multiple realisation: Keeping it real. In J. Hohwy & J. Kallestrup (Eds.), *Being reduced* (pp. 164–175). Oxford: Oxford University Press.

Aoki, S. K., Lillacci, G., Gupta, A., Baumschlager, A., Schweingruber, D., & Khammash, M. (2019). A universal biomolecular integral feedback controller for robust perfect adaptation. *Nature, 570,* 533–537.

Atmanspacher, H., & beim Graben, P. (2009). Contextual emergence. *Scholarpedia, 4*(3), 7997.

Bass, J., & Lazar, M. A. (2016). Circadian time signatures of fitness and disease. *Science, 354,* 994–999.

Bathgate, K. E., Bagley, J. R., Jo, E., Talmadge, R. J., Tobias, I. S., Brown, L. E., et al. (2018). Muscle health and performance in monozygotic twins with 30 years of discordant exercise habits. *European Journal of Applied Physiology*, *118*, 2097–2110.

Batterman, R. W. (2018). Autonomy of theories: An explanatory problem. *Nous*, *52*, 858–873.

Bechtel, W. (2018). The importance of constraints and control in biological mechanisms: Insights from cancer research. *Philosophy of Science*, *85*, 573–593.

Bedau, M. (2002). Downward causation and the autonomy of weak emergence. *Principia: An International Journal of Epistemology*, *6*, 5–50.

Berridge, M. (2014). *Cell signalling biology*. London: Portland Press.

Bickle, J. (2019). Multiple realizability. In E. N. Zalta (Ed.), *The Stanford Encyclopedia of Philosophy*.

Bishop, C. M. (1995). *Neural networks for pattern recognition*. Oxford: Oxford University Press.

Blachowicz, J. (2013). The constraint interpretation of physical emergence. *Journal for General Philosophy of Science*, *44*, 21–40.

Booch, G. (2006). *Object oriented analysis and design with application*. Boston: Pearson Education.

Boogerd, F. C., Bruggeman, F. J., Richardson, R. C., Stephan, A., & Westerhoff, H. V. (2005). Emergence and its place in nature: A case study of biochemical networks. *Synthese*, *145*, 131–164.

Briat, C., Gupta, A., & Khammash, M. (2019). Antithetic integral feedback ensures robust perfect adaptation in noisy biomolecular networks. *Cell Systems*, *2*, 15–26.

Bronowski, J. (2011). *The ascent of man*. New York: Random House.

Brooks, B. R., et al. (2009). CHARMM: The biomolecular simulation program. *Journal of Computational Chemistry*, *30*, 1545–1614.

Brooks, B. R., Bruccoleri, R. E., Olafson, B. D., States, D. J., Swaminathan, S. A., & Karplus, M. (1983). CHARMM: A program for macromolecular energy, minimization, and dynamics calculations. *Journal of Computational Chemistry*, *4*, 187–217.

Campbell, D. T. (1974). 'Downward causation' in hierarchically organised biological systems. In F. J. Ayala & T. Dobzhansky (Eds.), *Studies in the Philosophy of Biology* (pp. 179–186). London: Macmillan.

Campbell, N. A., & Reece, J. B. (2005). *Biology*. San Francisco: Benjamin Cummings.

Carroll, S. B. (2005). *The new science of Evo Devo—Endless forms most beautiful*. New York: Norton.

Catterall, W. A. (1995). Structure and function of voltage-gated ion channels. *Annual Review of Biochemistry*, *64*, 493–531.

Chalmers, D. J. (2006). Strong and weak emergence. In P. Clayton & P. C. W. Davies (Eds.), *The re-emergence of emergence* (pp. 244–256). Oxford: Oxford University Press.

Changeux, J. P., & Connes, A. (1998). *Conversations on mind, matter, and mathematics*. Princeton University Press.

Churchland, P. M. (2013). *Plato's camera: How the physical brain captures a landscape of abstract universals*. Cambridge: MIT Press.

D'Souza, A., Pearman, C. M., Wang, Y., Nakao, S., Logantha, S. J. R., Cox, C., et al. (2017). Targeting miR-423-5p reverses exercise training-induced HCN4 channel remodeling and sinus bradycardia. *Circulation Research*, *121*(9), 1058–1068.

Deacon, T. W. (1998). *The symbolic species: The co-evolution of language and the brain*. New York: Norton.

Dennett, D. C. (1996). *Darwin's dangerous idea*. London: Penguin.

Deweerdt, S. (2017). Sea change. *Nature*, *550*, S54–S58.

Donald, M. (2002). *A mind so rare*. Norton.

Drossel, B., & Ellis, G. F. R. (2018). Contextual Wavefunction Collapse: An integrated theory of quantum measurement. *New Journal of Physics*, *20*, 113025.

Drossel, B. (2019). *Strong emergence in condensed matter physics*. preprint. arXiv:1909.01134.

Dunbar, R. I. (2003). The social brain: Mind, language, and society in evolutionary perspective. *Annual Review of Anthropology*, *32*, 163–181.

Eddington, A. S. (1927). *The nature of the physicalworld*. Cambridge: Cambridge University Press (reprinted 2012).

Ellis, G. F. R. (2005). Physics, complexity and causality. *Nature, 435*, 743.

Ellis, G. F. R. (2014). The evolving block universe and the meshing together of times. *Annals of the New York Academy of Sciences, 1326*, 26–41. arXiv: 1407.7243.

Ellis, G. F. R., & Drossel, B. (2019). How downward causation occurs in digital computes. *Foundations of Physics, 49*, 1253–1277.

Ellis, G. F. R., & Drossel, B. (2020). Emergence of time. *Foundations of Physics, 50*, 161–190.

Ellis, G. F. R. & Kopel, J. (2018). The dynamical emergence of biology from physics: Branching causation via biomolecules. *Frontiers in Physiology, 9*, 1966.

Ellis, G. F. R., & Solms, M. (2017). *Beyond evolutionary psychology*. Cambridge: Cambridge University Press.

Feldman, G. (2019). Why neutrons and protons are modified inside nuclei. *Nature, 566*, 332–333.

Fink, M., & Noble, D. (2008). Noble model. *Scholarpedia, 3*(2), 1803.

Franklin, A., & Knox, E. (2018). Emergence without limits: The case of phonons. *Studies in History and Philosophy of Science Part B: Studies in History and Philosophy of Modern Physics, 64*, 68–78.

Fresco, N. (2012). The explanatory role of computation in cognitive science. *Minds and Machines, 22*, 353–380.

Frith, C. (2013). *Making up the mind: How the brain creates our mental world*. Hoboken: Wiley.

Ghirardi, G. (2007). *Sneaking a look at god's cards: Unraveling the mysteries of quantum mechanics*. Cambridge: Princeton University Press.

Gibb, S., Hendry, R. F., & Lancaster, T. (Eds.). (2019). *The Routledge handbook of emergence*. Oxfordshire: Routledge.

Gilbert, S. F. (2006). *Developmental biology*. Sunderland: Sinauer.

Gilbert, S. F., & Epel, D. (2009). *Ecological developmental biology*. Sunderland: Sinauer.

Gintis, H. (2011). Gene-culture coevolution and the nature of human sociality. *Philosophical Transactions of the Royal Society B: Biological Sciences, 366*, 878–888.

Glimcher, P. W. (2005). Indeterminacy in brain and behaviour. *Annual Review of Psychology, 56*, 25–56.

Godfrey-Smith, P. (2016). *Other minds: The octopus, the sea, and the deep origins of consciousness*. Straus and Giroux: Farrar.

Goelzer, A., Brikci, F. B., Martin-Verstraete, I., Noirot, P., Bessières, P., Aymerich, S., et al. (2008). Reconstruction and analysis of the genetic and metabolic regulatory networks of the central metabolism of Bacillus subtilis. *BMC Systems Biology, 2*(1), 2097–2110.

Grant, B. J., Gorfe, A. A., & McCammon, J. A. (2010). Large conformational changes in proteins: Signaling and other functions. *Current Opinion in Structural Biology, 20*, 142–147.

Grundmann, M. (2010). *Physics of semiconductors*. Berlin: Springer.

Guay, A., & Sartenaer, O. (2018). Emergent quasiparticles. Or how to get a rich physics from a sober metaphysics. In O. Bueno, R.-L. Chen, & M. B. Fagan (Eds.), *Individuation, process and scientific practices* (pp. 214–234). Oxford: OUP.

Harari, Y. N. (2014). *Sapiens: A brief history of humankind*. New York: Random House.

Hartwell, L. H., et al. (1999). From molecular to modular cell biology. *Nature, 402*, C47–C52.

Hasan, M. Z., & Kane, C. L. (2010). Colloquium: Topological insulators. *Reviews of Modern Physics, 82*(4), 3045.

Hendry, R. F. (2010). Emergence vs. reduction in chemistry. In C. MacDonald & G. MacDonald (Eds.), *Emergence in mind* (pp. 205–221). Oxford: Oxford University Press.

Hodges, A. (1992). *Alan Turing: The Enigma*. Vintage Books.

Hoffmann, P. (2012). *Life's Ratchet: How molecular machines extract order from Chaos*. New York: Basic Books.

Hofmeyr, J. H. S. (2017). Basic biological anticipation. In R. Poli (Ed.), *Handbook of anticipation*. Heidelberg: Springer.

Hofmeyr, J. H. S. (2018). Causation, constructors and codes. *Biosystems, 164*, 121–127.

Hordijk, W. (2013). Autocatalytic sets: from the origin of life to the economy. *BioScience, 63*, 877–881.

Humphreys, P. (1997). Emergence, not supervenience. *Philosophy of Science, 64*, S337–S345.

Humphreys, P. (2016). *Emergence: A philosophical account*. Oxford: Oxford University Press.

Jeffery, K., Pollack, R., & Rovelli, C. (2019). On the statistical mechanics of life: Schrödinger revisited. preprint. arXiv: 1908.08374.

Juarrero, A. (1999). *Dynamics in action: Intentional behavior as a complex system*. Cambridge: MIT Press.

Kandel, E. R. (1998). A new intellectual framework for psychiatry. *American Journal of Psychiatry, 155*, 457–469.

Kandel, E. R. (2001). The molecular biology of memory storage: A dialogue between genes and synapses. *Science, 294*(5544), 1030–1038.

Kandel, E. R. (2012). *The age of insight: The quest to understand the unconscious in art, mind, and brain, from Vienna 1900 to the present*. New York: Random House.

Kandel, E. R., Schwartz, J. H., Jessell, T. M., Siegelbaum, S. A., & Hudspeth, A. J. (2013). *Principles of neural science*. New York: McGraw Hill Professional.

Kanter, I., Kopelowitz, E., Vardi, R., Zigzag, M., Kinzel, W., Abeles, M., et al. (2011). Nonlocal mechanism for cluster synchronization in neural circuits. *EPL (Europhysics Letters), 93*(6), 66001.

Karplus, M. (2014). Development of multiscale models for complex chemical systems: From H+ H2 to biomolecules. *Angewandte Chemie International Edition, 53*, 9992–10005.

Keimer, B., & Moore, J. E. (2017). The physics of quantum materials. *Nature Physics, 13*, 1045.

Kim, J. (1998). *Mind in a physical world*. Cambridge: MIT Press.

Kim, J. (1999). Supervenient properties and micro-based properties: A reply to Noordhof. *Proceedings of the Aristotelian Society, 99*, 115–118.

King, R. A., Siddiqi, A., Allen, W. D., & Schaefer III, H. F. (2010). Chemistry as a function of the fine-structure constant and the electron-proton mass ratio. *Physical Review A, 81*(4), 042523.

Knuth, D. E. (1973). The art of computer programming: Vol. 1. Fundamental algorithms. Reading: Addison-Wesley.

Lancaster, T., & Pexton, M. (2015). Reduction and emergence in the fractional quantum Hall state. *Studies in History and Philosophy of Science Part B: Studies in History and Philosophy of Modern Physics, 52*, 343–357.

Laughlin, R. B. (1999). Fractional quantization. *Reviews of Modern Physics, 71*, 863.

Leggett, A. J. (1992). On the nature of research in condensed-state physics. *Foundations of Physics, 22*, 221–233.

Lehn, J.-M. (1993). Supramolecular chemistry. *Science, 260*, 1762–1764.

Lehn, J.-M. (1995). *Supramolecular chemistry*. Weinheim: VCH Verlagsgesellschaft mbH.

Lindholm, T., Yellin, F., Bracha, G., & Buckley, A. (2014). *The Java virtual machine specification*. Pearson Education.

Lucas, J. R. (1996). The unity of science without reductionism. *Acta Analytica, 15*, 89–95.

Luisi, P. L. (2002). Emergence in chemistry: Chemistry as the embodiment of emergence. *Foundations of Chemistry, 4*, 183–200.

MacCormick, J. (2011). *Nine algorithms that changed the future: The ingenious ideas that drive today's computers*. Cambridge: Princeton University Press.

Magleby, K. L. (2017). Ion-channel mechanisms revealed. *Nature, 541*, 33–34.

Mayr, E. (2002). *What evolution is*. Hurst: Phoenix.

McCleish, T. (2017). Strong emergence and downward causation in biological physics. *Philosophica, 92*, 113–138.

McLeish, T. (2019). Soft matter—An emerging interdisciplinary science of emergent entities. In S. Gibb, R. F. Hendry, & T. Lancaster (Eds.), *The Routledge handbook of emergence* (pp. 248–264). Milton Park: Routledge.

McLeish, T., Pexton, M., & Lancaster, T. (2019). Emergence and topological order in classical and quantum systems. *Studies in History and Philosophy of Science Part B: Studies in History and Philosophy of Modern Physics, 66*, 155–169.

Mellisinos, A. C. (1990). *Principles of modern technology*. Cambridge: Cambridge University Press.

Menzies, P. (2001). Counterfactual theories of causation. In *The Stanford Encyclopedia of Philosophy*.

Milo, R., Shen-Orr, S., Itzkovitz, S., Kashtan, N., Chklovskii, D., & Alon, U. (2002). Network motifs: Simple building blocks of complex networks. *Science, 298*, 824–827.

Montévil, M., & Mossio, M. (2015). Biological organisation as closure of constraints. *Journal of Theoretical Biology, 372*, 179–191.

Montévil, M., Mossio, M., Pocheville, A., & Longo, G. (2016). Theoretical principles for biology: Variation. *Progress in Biophysics and Molecular Biology, 122*(1), 36–50.

Mossio, M., Montévil, M., & Longo, G. (2016). Theoretical principles for biology: Organization. *Progress in Biophysics and Molecular Biology, 122*(1), 24–35.

Mossio, M., Saborido, C., & Moreno, A. (2009). An organizational account of biological functions. *The British Journal for the Philosophy of Science, 60*(4), 813–841.

Mossio, M., & Moreno, A. (2020). Organisational closure in biological organisms. *History and Philosophy of the Life Sciences, 32*(2/3), 269–88.

Müller, G. B. (2007). Evo-devo: Extending the evolutionary synthesis. *Nature Reviews Genetics, 8*(12), 943–949.

Natarajan, C., Hoffmann, F. G., Weber, R. E., Fago, A., Witt, C. C., & Storz, J. F. (2016). Predictable convergence in hemoglobin function has unpredictable molecular underpinnings. *Science, 354*(6310), 336–339.

Nichols, A. L. A., et al. (2017). A global brain state underlies C. elegans sleep behavior. *Science, 356.6344*, 1247.

Noble, D. (2002). Modeling the heart-from genes to cells to the whole organ. *Science, 295*, 1678–1682.

Noble, D. (2008). *The music of life: Biology beyond genes*. Oxford: Oxford University Press.

Noble, D. (2011). A theory of biological relativity: No privileged level of causation. *Interface Focus, 2*, 55–64.

Noble, D. (2017). Evolution viewed from physics, physiology and medicine. *Interface Focus, 7*(5), 20160159.

Noble, R., & Noble, D. (2018). Harnessing stochasticity: How do organisms make choices? *Chaos: An Interdisciplinary Journal of Nonlinear Science, 28*(10), 106309.

Noble, R., Tasaki, K., Noble, P. J., & Noble, D. (2019). Biological relativity requires circular causality but not symmetry of causation: So, where, what and when are the boundaries? *Frontiers in Physiology, 10*, 827.

Nurse, P. (2008). Life, logic and information. *Nature, 454*, 424.

O'Connor, T., & Wong, H. Y. (2015). Emergent properties. In E. N. Zalta (Ed.), *The Stanford Encyclopedia of Philosophy*.

O'Gorman, T. J., Ross, J. M., Taber, A. H., Ziegler, J. F., Muhlfeld, H. P., Montrose, C. J., et al. (1996). Field testing for cosmic ray soft errors in semiconductor memories. *IBM Journal of Research and Development, 40*, 41–50.

Oyama, S., Griffiths, P. E., & Gray, R. D. (2001). *Cycles of contingency: Developmental systems and evolution*. Cambridge: MIT Press.

Passon, O. (2019). *Completeness and quantum theory: From the spectral gap to EPR and back again*. Lecture notes, Bergisches Universität Wuppertal.

Penrose, R. (2000). *The large, the small and the human mind*. Cambridge: Cambridge University Press.

Percival, I. (1991). Schrödinger's quantum cat. *Nature, 351*, 357.

Peter, I. S., & Davidson, E. H. (2011). Evolution of gene regulatory networks controlling body plan development. *Cell, 144*(6), 970–985.

Peter, P., & Uzan, J. P. (2013). *Primordial cosmology*. Oxford: Oxford University Press.

Petsko, G. A., & Ringe, D. (2009). *Protein structure and function*. Oxford: Oxford University Press.

Phillips, P. (2012). *Advanced solid state physics*. Cambridge: Cambridge University Press.

Pigliucci, M., & Müller, G. B. (2000). *Evolution—The extended synthesis*. Cambridge: MIT Press.

Qi, X.-L., & Zhan, S.-C. (2011). Topological insulators and superconductors. *Reviews of Modern Physics*, *83*, 1057.

Randall, D., Burggren, W., & French, K. (2002). *Eckert animal physiology: Mechanisms and adaptations*. New York: W. H. Freeman.

Ranjan, R., Logette, E., Marani, M., Herzog, M., Tache, V., & Markram, H. (2019). A kinetic map of the homomeric voltage-gated potassium channel (Kv) family. *Frontiers in Cellular Neuroscience*, *13*, 358.

Ravasz, E., Somera, A. L., Mongru, D. A., Oltvai, Z. N., & Barabási, A. L. (2002). Hierarchical organization of modularity in metabolic networks. *Science*, *297*, 1551–1555.

Rhoades, R., & Pflanzer, R. (1989). *Human physiology*. Fort Worth: Saunders College Publishing.

Rolls, E. T., & Deco, G. (2010). *The noisy brain: Stochastic dynamics as a principle of brain function*. Oxford: Oxford University Press.

Rosen, R. (1958). A relational theory of biological systems. *The Bulletin of Mathematical Biophysics*, *20*, 245–260.

Sales-Pardo, M. (2017). The importance of being modular. *Science*, *357*, 128–129.

Santo, D. (2019). *Flavio, and Nicolas Gisin. "Physics without determinism: Alternative interpretations of classical physics.".* Phys. Rev. A, 100, 062107. preprint. arXiv: 1909.03697.

Sartenaer, O. (2015). Synchronic vs. diachronic emergence: A reappraisal. *European Journal for Philosophy of Science*, *5*, 31–54.

Sauro, H. M. (2017). Control and regulation of pathways via negative feedback. *Journal of the Royal Society Interface*, *14*(127), 20160848.

Scalo, J., Wheeler, J. C., & Williams, P. (2001). Intermittent jolts of galactic UV radiation: Mutagenic effects. In L. M. Celnekier (Ed.), *Frontiers of life, 12th Recontres de Blois*. preprint. astro-ph/0104209.

Scott, A. (1999). *Stairway to the mind: The controversial new science of consciousness*. Berlin: Springer Science and Business Media. .

Simon, H. A. (1996). *The architecture of complexity. Sciences of the artificial* (3. edn.). Cambridge: MIT Press.

Simon, S. H. (2013). *The Oxford solid state basics*. Oxford: Oxford University Press.

Tanenbaum, A. S. (2006). *Structured computer organisation*. Englewood Cliffs: Prentice Hall.

Thompson, C. (2019). *Coders: Who they are, what they think, and how they are changing the world*. London: Picador.

Tokura, Y., Kawasaki, M., & Nagaosa, N. (2017). Emergent functions of quantum materials. *Nature Physics*, *13*, 1056.

Ulanowicz, R. E. (1995). Utricularia's secret: The advantage of positive feedback in oligotrophic environments. *Ecological Modelling*, *79*, 49–57.

Unwin, N. (1993). Neurotransmitter action: Opening of ligand-gated ion channels. *Cell*, *72*, 31–41.

Venema, L., Verberck, B., Georgescu, I., Prando, G., Couderc, E., Milana, S., et al. (2016). The quasiparticle zoo. *Nature Physics*, *12*(12), 1085–1089.

Wagner, A. (2011). *The origins of evolutionary innovations*. Oxford: Oxford University Press.

Wagner, A. (2017). *Arrival of the fittest*. New York: Penguin Random House.

Walker, S. I., Kim, H., & Davies, P. C. (2016). The informational architecture of the cell. *Philosophical Transactions of the Royal Society*, *374*, 20150057.

Wang, H. (1997). *A logical journey from Gödel to philosophy*. Cambridge: MIT Press.

Watson, J. D., Bell, S. P., Gann, A., Levine, M., Losick, R., & Baker, T. A. (2013). *Molecular biology of the gene*. Pearson.

Weinberg, S. (1995). *The quantum theory of fields foundations* (Vol. 1). Cambridge: Cambridge University Press.

West-Eberhard, M. J. (2003). *Developmental plasticity and evolution*. New York: Oxford University Press.

Winning, J., & Bechtel, W. (2018). Rethinking causality in biological and neural mechanisms: Constraints and control. *Minds and Machines*, *28*, 287–310.

Wolpert, L. (2002). *Principles of development*. Oxford: Oxford University Press.

Wu, Z. S., Cheng, H., Jiang, Y., Melcher, K., & Xu, H. E. (2015). Ion channels gated by acetylcholine and serotonin: Structures, biology, and drug discovery. *Acta Pharmacologica Sinica, 36,* 895.

Ziegler, J. F., & Lanford, W. A. (1979). Effect of cosmic rays on computer memories. *Science, 206,* 776–788.

Physik ohne Reduktion

Rico Gutschmidt

Nach dem vorherrschenden reduktionistischen Selbstverständnis der Physik sind höherstufige physikalische Theorien als Grenzfall in grundlegenderen Theorien enthalten und daher im Prinzip überflüssig. Höherstufige Theorien werden laut dieser Sichtweise nur aus dem pragmatischen Grund verwendet, für ihren jeweiligen Bereich besser anwendbar zu sein. Es gibt zwar verschiedene Modelle der Reduktion physikalischer Theorien, dieses Selbstverständnis lässt sich aber am besten mit dem Modell der eliminativen Reduktion auf den Begriff bringen, nach dem eine Theorie auf eine andere reduziert wird, wenn sie im Prinzip nicht mehr benötigt wird, da alle von ihr erklärten Phänomene auch von der reduzierenden Theorie erklärt werden. Laut diesem Reduktionskonzept darf die reduzierende Theorie bei ihren Erklärungen nicht auf die Begriffe, Konzepte und Modelle der reduzierten Theorie zurückgreifen, da diese sonst immer noch gebraucht würde und nicht überflüssig wäre. Obwohl das reduktionistische Selbstverständnis der Physik auf diesen eliminativen Reduktionsbegriff hinausläuft, ist es sich über diesen naheliegenden Punkt offenbar nicht im Klaren, da er zu einer Frage führt, die die reduktionistische Position beantworten müsste, aus prinzipiellen Gründen aber nicht beantworten kann. Diese Frage wird im Folgenden entwickelt. Der besondere Witz dieser Frage wird darin bestehen, dass sie keine Phänomene oder Theorien ins Spiel bringt, die in der Fachwelt umstritten oder nicht allgemein anerkannt sind, sondern auf einer ganz basalen Reflexion beruht, die das Modell der eliminativen Reduktion auf den etablierten Stand der Physik bezieht.

Der erste Schritt auf dem Weg zu dieser Frage besteht in einem genaueren Blick auf die Redeweise, eine Theorie sei in einer anderen als Grenzfall enthalten. Diese Redeweise suggeriert, eine solche Grenzfalltheorie sei in obigem Sinne überflüssig, da eine Erklärung mit dieser Theorie im Prinzip auf eine Erklärung mit der grundlegenderen Theorie hinausläuft. Dagegen spricht aber der einfache Punkt, dass sich die Vorhersagen der beteiligten Theorien in allen interessanten Fällen widersprechen,

R. Gutschmidt (✉)
Universität Konstanz, Konstanz, Deutschland

© Der/die Autor(en), exklusiv lizenziert durch Springer-Verlag GmbH, DE, ein Teil von Springer Nature 2021
O. Passon und C. Benzmüller (Hrsg.), *Wider den Reduktionismus,*
https://doi.org/10.1007/978-3-662-63187-4_7

weshalb es sich bei einem solchen Grenzfall nicht um eine logische Ableitung der einen Theorie aus der anderen Theorie handeln kann, sondern eher um eine näherungsweise Herleitung. So fallen etwa nach dem Galileischen Fallgesetz Körper mit konstanter Beschleunigung auf die Erde, während diese Beschleunigung nach dem Newtonschen Gravitationsgesetz mit kleinerem Abstand zur Erde zunimmt. Nach Newton ist die Beschleunigung erst konstant, wenn der Körper bereits auf der Erde liegt, während das Galileische Gesetz der konstanten Beschleunigung für fallende Körper gilt. Dies ist ein klarer Widerspruch in den Vorhersagen der beiden Theorien, die daher nicht logisch ineinander enthalten sein können, sondern in einer Näherungsrelation stehen, nach der für kleine Abstände zur Erdoberfläche aus der Newtonschen Perspektive näherungsweise das Galileische Fallgesetz gilt. Der entscheidende Punkt an dieser Überlegung besteht nun darin, dass es sich bei einer solchen näherungsweisen Herleitung ohne logische Ableitung um eine Näherungsrelation zwischen zwei unabhängigen und eigenständigen Theorien handelt. Über die besagten Widersprüche hinaus haben physikalische Theorien ihre jeweiligen Grundannahmen und Modelle, von denen man zwar zeigen kann, dass sie in bestimmten Bereichen ähnliche Vorhersagen machen, was aber, trotz der üblichen Redeweise, nicht bedeutet, dass eine Theorie in der anderen enthalten ist. Wenn eine Theorie tatsächlich in einer anderen Theorie enthalten wäre, würden sich die Theorien nicht nur nicht widersprechen, sondern auch dieselben Grundannahmen und Modelle verwenden, da sich nur so eine logische Relation etablieren lässt. Das ist aber in allen interessanten Beispielen nicht der Fall. So beruht das Newtonsche Gravitationsgesetz auf dem Konzept der Kraft, das es im Galileischen Ansatz nicht gibt. Trotz der näherungsweisen Herleitung ist daher das Galileische Fallgesetz nicht in der Newtonschen Theorie enthalten, sondern bleibt ein eigenständiges Gesetz, das für Körper in der Nähe der Erdoberfläche eine ähnliche Vorhersage macht wie das Newtonsche Gravitationsgesetz. Das ist eine geradezu triviale Feststellung, die auch kein Reduktionist bestreiten würde. Dennoch hat sie weitreichende Konsequenzen für eine reduktionistische Position, wie sich im Folgenden zeigen wird.

So besteht der zweite Schritt auf dem Weg zu der unbeantwortbaren Frage in der Folgerung, dass höherstufige Theorien nicht deshalb überflüssig sind, weil sie in grundlegenderen Theorien logisch enthalten wären, was sie, wie gezeigt, nicht sind, sondern weil die grundlegenderen Theorien die Phänomene erklären können, die von den höherstufigen Theorien erklärt werden. Dabei sind die Erklärungen der grundlegenderen Theorien üblicherweise auch besser als die der höherstufigen, was sich nicht zuletzt darin zeigt, dass der Grad der Abweichung zu den Erklärungen der höherstufigen Theorie angegeben werden kann. Wenn die reduktionistische Position also über das Bestehen einer näherungsweisen Herleitung hinaus behauptet, dass höherstufige Theorien durch grundlegendere Theorien überflüssig geworden sind, kann sie nach dem ersten Schritt der Argumentation nicht darauf verweisen, dass diese in letzteren enthalten seien, sondern muss im Sinne des Modells der eliminativen Reduktion die Frage beantworten, wie die grundlegenderen Theorien alle Phänomene erklären können, die von der jeweiligen höherstufigen Theorie erklärt werden, und zwar ohne dabei Konzepte der höherstufigen Theorie zu verwenden, da diese sonst nicht überflüssig wäre. Auch dies ist eine geradezu triviale Forderung, die aber im Kern bereits die gesuchte unbeantwortbare Frage enthält: Solange eine Theorie zur Erklärung von physikalischen Phänomenen noch benötigt wird, ist sie trotz näherungsweiser Herleitung aus einer anderen Theorie nicht überflüssig.

Im dritten Schritt der Argumentation wird nun entsprechend gezeigt, dass diese trivial erscheinende Forderung nicht in jedem Fall erfüllt werden kann, da zahlreiche physikalische Phänomene nach wie vor von den jeweils zuständigen höherstufigen Theorien erklärt werden, ohne dass eine Erklärung allein durch eine grundlegendere Theorie vorliegt. Im Kontext der Gravitationsphysik erklärt zwar das Newtonsche Gravitationsgesetz das Phänomen fallender Körper, und zwar sogar besser als das Galileische Fallgesetz, womit dieses tatsächlich überflüssig ist. Für das Verhältnis zwischen dem Newtonschen Gravitationsgesetz und der Allgemeinen Relativitätstheorie ist die Sache aber nicht so einfach. Zunächst können zwar Näherungsbeziehungen zwischen den Gesetzen dieser Theorien entwickelt werden, bei denen es sich aber nicht um logische Ableitungen handelt. Neben den Widersprüchen in den Vorhersagen sind die Unterschiede in den Grundannahmen und Modellen der Theorien so fundamental, dass sie nicht im Sinne der Logik überbrückt werden können. Im Rahmen solcher Näherungsbeziehungen werden etwa aus den Gesetzen der Allgemeinen Relativitätstheorie Konzepte entwickelt, die unter bestimmten Bedingungen denen der Newtonschen Theorie ähneln, aber nie ganz entsprechen. Es wird zum Beispiel argumentiert, dass im Grenzfall einer leeren Raumzeit ohne Massen die Bedingungen des Newtonschen Modells gelten, wobei letzteres aber Interaktionen zwischen echten Massen beschreibt, die es in einer leeren Raumzeit gerade nicht gibt. Wie im Verhältnis zwischen Galileis Fallgesetz und der Newtonschen Gravitation gibt es zwar Näherungsbeziehungen, aber keine logische Ableitung; es handelt sich um zwei eigenständige und unabhängige Theorien, die nicht logisch ineinander enthalten sind. Während aber die Newtonsche Theorie ebenfalls fallende Körper beschreibt und damit das Galileische Fallgesetz tatsächlich überflüssig macht, gibt es zahlreiche Gravitationsphänomene, die von der Newtonschen Theorie erklärt werden, aber nicht von der Allgemeinen Relativitätstheorie.

So wird die Interaktion zwischen den Planeten im Sonnensystem trotz Allgemeiner Relativitätstheorie mit dem Newtonschen Gravitationsgesetz beschrieben, da die Feldgleichungen für eine derart komplexe Konstellation nicht ohne Weiteres gelöst werden können. Mit der Schwarzschildlösung wird zwar eine kleine Anomalie in der Bahn des Merkur erklärt. Die für die Planetenbahnen viel maßgeblicheren Interaktionen zwischen den Planeten lassen sich aber mit der Schwarzschild-Lösung aus prinzipiellen Gründen nicht beschreiben, da sie von einem Universum mit einer einzigen, zentralen Masse ausgeht und die Planeten als masselose Probeteilchen ansieht. Die Interaktionen zwischen den tatsächlichen Massen der Planeten kann so prinzipiell nicht erfasst werden, und sie werden auch nach wie vor mit der Newtonschen Physik beschrieben. Dabei kann man nun aber nicht sagen, es würde sich wegen der Grenzfallrelation letztlich um eine Erklärung der Allgemeinen Relativitätstheorie handeln, da diese Relation, wie gezeigt, bei näherem Hinsehen eine

Näherungsbeziehung zwischen zwei eigenständigen und unabhängigen Theorien darstellt. Die Newtonsche Theorie ist also trotz dieser Näherungsbeziehung ganz offensichtlich nicht überflüssig. Dies wäre erst der Fall, wenn die Interaktionen zwischen den Planeten allein mit den Konzepten der Allgemeinen Relativitätstheorie beschrieben werden könnten, ohne dabei die Konzepte der Newtonschen Theorie zu verwenden. So müssten die Feldgleichungen für das gesamte Sonnensystem mit allen einzelnen Massen aufgestellt und gelöst werden. Ähnliches gilt für Doppelsternsysteme, Sternhaufen, oder Galaxien bzw. Galaxiencluster. All dies sind Phänomene, die von der Newtonschen Theorie beschrieben werden, nicht aber von der Allgemeinen Relativitätstheorie. Das bedeutet natürlich nicht, dass die Feldgleichungen für diese komplexen Konstellationen nicht gültig wären; die Allgemeine Relativitätstheorie ist ohne Frage die bessere Gravitationstheorie. Es kann aber sogenannte emergente Phänomene geben, die zwar den Feldgleichungen genügen, aufgrund ihrer Komplexität aber nicht ohne Zuhilfenahme geeigneter höherstufiger Theorien beschrieben werden können. Wenn man also behauptet, die Newtonsche Gravitationstheorie sei im Prinzip überflüssig, müsste man solche emergenten Phänomene ausschließen, indem man zeigt, wie sich alle die genannten komplexen Konstellationen allein in Begriffen der Allgemeinen Relativitätstheorie beschreiben lassen. Für die Physik sind solche Beschreibungen nicht notwendig, da alle diese Phänomene mit der Newtonschen Physik beschrieben werden, was aufgrund der besagten Näherungsbeziehungen auch gut gerechtfertigt ist. Es geht hier lediglich um die prinzipielle Frage, ob die Newtonsche Gravitationstheorie nicht mehr gebraucht wird.

In dieser Frage ist es nun naheliegend zu argumentieren, dass sich die Feldgleichungen im Prinzip für alle die genannten komplexen Konstellationen aufstellen und numerisch lösen lassen. Selbst wenn solche Computersimulationen die gegenwärtigen Rechnerkapazitäten deutlich übersteigen, scheint man doch sagen zu können, dass die genannten, anscheinend emergenten Phänomene prinzipiell numerisch mit den Feldgleichungen beschrieben werden können. Demnach würde man die Newtonsche Gravitationstheorie nur aus pragmatischen Gründen verwenden, im Prinzip wäre sie überflüssig. Gegen diese Auffassung lassen sich aber durchaus Argumente ins Spiel bringen. So muss jede numerische Simulation Startwerte voraussetzen, die man nicht ohne Zirkelschluss aus der Simulation selbst gewinnen kann. Bei den Computermodellen verschmelzender Schwarzer Löcher zum Beispiel beginnt die Simulation in einer Entfernung, die sich (post-)newtonsch beschreiben lässt. Da sich entsprechende Startwerte aufgrund der Komplexität nicht aus den Feldgleichungen ableiten lassen, wird die Newtonsche Theorie also auch bei einer numerischen Lösung der Feldgleichungen in einem eminenten Sinne nach wie vor benötigt. Außerdem müssen die Zahlenwerte einer numerischen Simulation physikalisch interpretiert werden, wozu im Falle der Simulation verschmelzender Schwarzer Löcher wiederum Konzepte der Newtonschen Theorie herangezogen werden. Dies gilt übrigens bereits für die exakte Schwarzschild-Lösung, bei der sich eine Integrationskonstante nur durch den Vergleich mit dem Newtonschen Grenzfall als zentrale Masse interpretieren lässt. Die Probleme der Startwerte und der Interpretation sprechen damit gegen die Behauptung, die Newtonsche Theorie sei im Prinzip überflüssig. Allerdings lässt sich die Möglichkeit, die besagten komplexen Konstellationen dennoch allein mit den Mitteln der Allgemeinen Relativitätstheorie zu beschreiben, nicht absolut ausschließen. Auch wenn die aufgeführten Argumente dagegen sprechen und es derzeit nicht danach aussieht, kann der Reduktionist auf zukünftige Rechnerkapazitäten und numerische Methoden verweisen, die vielleicht doch eine Erklärung sämtlicher von der Newtonschen Physik beschriebener Gravitationsphänomene in Begriffen der Allgemeinen Relativitätstheorie ermöglichen und damit die Newtonsche Gravitationstheorie im hier beschriebenen Sinne überflüssig machen würden.

Im vierten und letzten Schritt der Argumentation wird daher nun gezeigt, dass die entsprechende Frage im Fall des Verhältnisses zwischen Newtonscher Mechanik und Quantenmechanik auch im Prinzip nicht beantwortet werden kann: Wie lässt sich die makroskopische Welt in Begriffen der Quantenmechanik beschreiben, ohne dabei Newtonsche Konzepte zu verwenden? Dabei handelt sich um eine physikalisch nicht motivierte Frage, da solche Beschreibungen nicht notwendig sind, schließlich gibt es ja für den makroskopischen Bereich die Newtonsche Theorie. Eine reduktionistische Position aber, die dem Selbstverständnis der Physik entspricht, nach dem höherstufige physikalische Theorien als Grenzfall in grundlegenderen Theorien enthalten und daher im Prinzip überflüssig sind, muss diese Frage beantworten können. Mit dem Verweis auf die Grenzfallrelation lässt sie sich nicht beantworten, da es sich auch bei dem Verhältnis zwischen diesen Theorien um Näherungsbeziehungen handelt und nicht um eine logische Ableitung. Wenn etwa im Rahmen des Ehrenfest-Theorems bestimmte strukturelle Analogien zwischen Newtonscher Mechanik und Quantenmechanik herausgestellt werden, handelt es sich nur um Analogien zwischen ansonsten eigenständigen und unabhängigen Theorien mit fundamental verschiedenen Grundannahmen und Modellen. Außerdem sind die Widersprüche zwischen den Vorhersagen dieser Theorien so grundlegend, dass trotz Analogien und Näherungsbeziehungen von einer logischen Ableitung nicht die Rede sein kann. Wenn die Newtonsche Physik also aufgrund der Quantenmechanik überflüssig sein soll, müssen die makroskopischen Phänomene der klassischen Mechanik quantenmechanisch erklärt werden können, ohne dabei die Newtonsche Theorie zu verwenden. In solchen Erklärungen müsste die Schrödinger-Gleichung für alle beteiligten Teilchen aufgestellt und gelöst werden, was zwar aufgrund der schieren Komplexität auch mit zukünftigen Rechnerkapazitäten unmöglich zu sein scheint, im Prinzip aber denkbar ist. Die Schrödinger-Gleichung lässt sich allerdings nur in einfachen Fällen wie dem Wasserstoffatom exakt lösen, bereits bei größeren Atomen und bei Molekülen ist man auf numerische Lösungen angewiesen, von wirklich makroskopischen Phänomenen ganz zu schweigen. Auf der anderen Seite gibt es makroskopische Phänomene, wie etwa turbulente Strömungen, die selbst aus der Perspektive der klassischen Mechanik zu komplex sind, um vollständig verstanden zu werden. Aus der Perspektive der Quantenmechanik ist die Komplexität solcher Phänomene ungleich größer, weshalb sie kaum etwas zu ihrem Verständnis beitragen dürfte. Dennoch lässt sich natürlich argumentieren, die komplexen und anscheinend emergenten Phänomene der makroskopischen Welt könnten im Prinzip ohne Zuhilfenahme der Newtonschen Physik mit numerischen Simulationen auf Grundlage der Quantenmechanik beschrieben werden. Dagegen spricht auch hier wieder das Problem der Startwerte und der Interpretation numerischer Zahlenwerte. So werden bereits im mikroskopischen Bereich

einzelner Atome und kleiner Moleküle nicht nur halbklassische Methoden einge-
setzt, sondern sogar klassische Zustände zugrunde gelegt. In der Quantenchemie wird
schließlich höherstufiges Wissen über die Struktur von Molekülen verwendet, um
die Zahlenwerte numerischer Lösungen der Schrödinger-Gleichung interpretieren zu
können, womit man bereits weit vor wirklich makroskopischen Phänomenen auf eine
höherstufige Theorie angewiesen ist. Wie im Fall der Feldgleichungen lässt sich aber
trotz des systematischen Problems der Interpretation nicht ausschließen, dass eines
Tages vielleicht doch größere Rechnerkapazitäten und entsprechende numerische
Methoden entwickelt werden, die eine Beschreibung auch komplexer makroskopischer Phänomene mit der Schrödinger-Gleichung ermöglichen.

Es gibt im Fall der Quantenmechanik allerdings ein weiteres Problem, das der
Reduktionist auch nicht mit Verweis auf die Zukunft beheben kann. Wenn man
nämlich tatsächlich ein makroskopisches Phänomen quantenmechanisch mit der
Schrödinger-Gleichung beschreiben könnte, müsste es sich gemäß dieser Theorie wie
Schrödingers Katze in einem Superpositionszustand befinden. Abgesehen von sehr
speziellen Fällen, wie etwa der Verschränkung zwischen räumlich getrennten atoma-
ren Gasen oder der angestrebten Überlagerungszustände in Quantencomputern, gibt
es aber in der makroskopischen Welt keine solchen Superpositionszustände, weshalb
die Quantenmechanik jenseits der besagten numerischen Probleme aus prinzipiellen
Gründen für Erklärungen im allgemeinen makroskopischen Phänomenbereich nicht
geeignet ist. Mit dem Konzept der Dekohärenz wird zwar argumentiert, dass sich
makroskopische Phänomene nur scheinbar in klassischen Zuständen befinden und in
Wirklichkeit den Gesetzen der Quantenmechanik genügen. Dies zeigt aber lediglich,
dass die Quantenmechanik in gewisser Weise auch für makroskopische Phänomene
gültig ist, was noch nicht bedeutet, dass sie diese Phänomene eigenständig, also
ohne Newtonsche Physik, beschreiben kann. Hierzu müsste man, wie gezeigt, die
Schrödinger-Gleichung für alle beteiligten Teilchen aufstellen, was durch den Erklä-
rungsansatz der Dekohärenz im Gegenteil noch komplizierter wird. Das Phänomen
der Dekohärenz beruht nämlich auf Wechselwirkungen mit der Umgebung, wobei
aber nicht klar ist, wo die Umgebung aufhört, weshalb man für die quantenmechani-
sche Beschreibung eines makroskopischen Phänomens die Schrödinger-Gleichung
nicht nur für das fragliche Phänomen und seine Umgebung, sondern letztlich für das
gesamte Universum aufstellen müsste. Dies ist nicht nur numerisch unmöglich, sondern nach dem Dekohärenzansatz auch physikalisch fragwürdig, da es für das ganze
Universum keine Umgebung mehr gibt und es sich also insgesamt in einem quantenmechanischen Zustand befinden müsste. Das Konzept der Dekohärenz kann damit
den hier vorgebrachten prinzipiellen Einwand gegen die Möglichkeit der Erklärung
makroskopischer Phänomene allein mit Begriffen der Quantenmechanik nicht ent-
kräften. Eine Erklärung allein mit Begriffen der Quantenmechanik impliziert, dass
sich die jeweiligen Phänomene in quantenmechanischen Zuständen befinden, die
zwar makroskopisch relevant sein können, wie etwa bei Supraleitung, Lasern oder
Neutronensternen, die es aber, von wenigen Spezialfällen abgesehen, in der makroskopischen Welt nicht gibt; und was es schließlich bedeuten würde, dass sich das
ganze Universum in einem quantenmechanischen Zustand befände, ist physikalisch
nicht klar. Die Frage, wie sich die makroskopische Welt in Begriffen der Quan-

tenmechanik erklären lässt, ohne dabei Newtonsche Konzepte zu verwenden, lässt sich daher prinzipiell nicht beantworten, auch nicht mit Verweis auf die Zukunft. Auch wenn nicht ganz klar ist, wie der Übergang zwischen der mikroskopischen und der makroskopischen Ebene verstanden werden kann, ist die Quantenmechanik keine geeignete Theorie für den makroskopischen Bereich und macht insbesondere die Newtonsche Theorie, trotz der bestehenden Näherungsbeziehungen, nicht überflüssig, auch nicht im Prinzip. Dies gilt analog für jede zukünftige „Theorie von Allem", die die Kluft zwischen quantenmechanischer mikroskopischer und klassischer makroskopischer Welt nicht überbrücken dürfte.

Die hiermit vorgelegte unbeantwortbare Frage widerlegt das vorherrschende reduktionistische Selbstverständnis der Physik. Gegen dieses Selbstverständnis zeigen die vier Schritte der vorgetragenen Argumentation, dass die Physik pluralistisch mit verschiedenen, eigenständigen Theorien auf verschiedenen Beschreibungsebenen arbeitet, die zwar fruchtbar miteinander kooperieren können, aber nicht im Sinne einer eliminativen Reduktion in grundlegenderen Theorien enthalten sind. Die Möglichkeit einer solchen Reduktion wurde im Fall der Newtonschen bzw. relativistischen Gravitationsphysik mit schwerwiegenden Argumenten infrage gestellt und für das Verhältnis zwischen klassischer Mechanik und Quantenmechanik ausgeschlossen. Insgesamt wurde damit gezeigt, dass das reduktionistische Selbstverständnis der Physik nicht zu ihrer pluralistischen Praxis passt, für die es sicher förderlich wäre, wenn sich die Physik in diesem Sinne selbst besser verstünde.[1]

---

[1] Ausführliche wissenschaftliche Ausarbeitungen der hier nur essayistisch vorgestellten Thesen finden sich in Gutschmidt (2009) *Einheit ohne Fundament. Eine Studie zum Reduktionsproblem in der Physik.* De Gruyter, Berlin. Vgl. auch Gutschmidt (2014) Reduction and the Neighbourhood of Theories: A New Approach to the Intertheoretic Relations in Physics. *J Gen Philos Sci* **45**(1): 49–70.

Teil III
Logisch-mathematische Perspektiven

Is there an Axiom for Everything?

8

Jean-Yves Béziau

We first start by clarifying what axiomatizing everything can mean. We then study a famous case of axiomatization, the axiomatization of natural numbers, where two different aspects of axiomatization show up, the model-theoretical one and the proof-theoretical one. After that we discuss a case of axiomatization in a sense opposed to the one of arithmetic, the axiomatization of the notion of order, where the idea is not to catch a specific structure, but a notion. A third mathematical case is then examined, the one of identity, a simple and obvious notion, but that cannot be axiomatized in first-order logic. We then move on to more general notions: the axiomatization of causality and the universe. To end with we deal with an even more tricky question: the axiomatization of reasoning itself. In conclusion we discuss in the light of our investigations the relation between axiomatization and understanding.

8.1 Axiomatizing Everything

One may want to axiomatize everything. Is it possible? To answer this question we need to understand what it means. This can be understood in two different ways:

1. Given *anything,* to axiomatize it.
2. A single axiomatization for the *whole thing.*

(1) is weaker than (2) and can be seen as a particular case. "Single axiomatization" is ambiguous; the extreme case is *one* axiom describing everything, the world, the whole reality, like a fundamental equation explaining the universe.

J.-Y. Béziau (✉)
Brazilian Academy of Philosophy, Brazilian Research Center, University of Brazil, Rio de Janeiro, Brasilien
E-mail: jyb@ufrj.br

© Der/die Autor(en), exklusiv lizenziert durch Springer-Verlag GmbH, DE, ein Teil von
Springer Nature 2021
O. Passon und C. Benzmüller (Hrsg.), *Wider den Reduktionismus,*
https://doi.org/10.1007/978-3-662-63187-4_8

In this paper we will examine if (1) and (2) are possible or not. To do that we need to understand what we have on both sides: *Axiom* and *Everything*. But the clue is the relation between the two, which can be qualified as *axiomatizing*. Although this notion is well-known, especially through the promotion of the so-called "Axiomatic Method" (see e.g. Gonseth 1939; Hintikka 2011), there are many confusions surrounding it. As often, the confusion is due to the mixture of different meanings attached to one word. Alfred Tarski analyzes the situation in the following way:

> We should reconcile ourselves with the fact that we are confronted, not with one concept, but with several different concepts which are denoted by one word; we should try to make these concepts as clear as possible (by means of definition, or of an axiomatic procedure, or in some other way); to avoid further confusions, we should agree to use different terms for different concepts; and then we may proceed to a quiet and systematic study of all concepts involved, which will exhibit their main properties and mutual relations (Tarski 1944).

This is here in relation with the concept of truth, and there are indeed some axiomatic theories of truth helping to clarify this concept (see Halbach 2011). Regarding the concept of axiomatization, it is not clear if we can axiomatize it, but at least we can give definitions of it.

It is good to start with a very general view, encompassing the different possible understandings, before making specific distinctions. We can say that *axiomatizing is finding some simple and obvious truths from which it is possible to master a certain field.* It is a way to concentrate the understanding of a field in a few statements, using some basic notions and properties that can grasp the rest. It can be seen as a kind of reduction, but not necessarily a negative reduction, a reductionism. We can metaphorically compare that to condensation or to be at the top of a mountain from where it is possible to see a whole region.

The axiomatic method started in Ancient Greece but it was developed in a much more sophisticated way in modern logic. To properly understand how it works, one needs to have a basic knowledge of the four theories forming the basis of modern logic: set theory, proof theory, recursion theory and model theory. We will here explain the axiomatic method from this perspective. But since our paper is for a wide audience and is rather philosophical we will not give too many technical details. Nevertheless what we are saying can be precisely developed at a technical level and we are giving precise references supporting what we are saying.

8.2   Axiomatizing Natural Numbers

Let us start with a basic, central and critical example: the axiomatization of the natural numbers. By contrast to geometry, arithmetic was axiomatized only at the end of the 19th century. This was done independently and in different ways by Peirce (1881), Dedekind (1888) and Peano (1889).

We have an intuitive idea of the natural numbers since childhood through their names, enumerating them: 0, 1, 2, 3 …, making operations with them and using them to order things. In modern mathematics natural numbers are considered as forming a structure,[1] which can be presented in different ways. The following one is quite close to our intuition about them:

$$\mathcal{N} = \langle \mathbb{N}; <, +, \times \rangle.$$

It is a set with a binary relation and two binary functions. The structure of natural numbers can be considered in other ways, for example adding a unary function of succession and a constant for the number zero:

$$\mathcal{N} = \langle \mathbb{N}; 0, s, <, +, \times \rangle.$$

From a *model-theoretical* point of view, axiomatizing the natural numbers means finding a set of axioms that characterizes the structure $\mathcal{N}$ in the sense that this structure is the only structure which verifies, obeys, is a *model* of these axioms. Here "only" means up to isomorphism. And since isomorphism depends on one-to-one correspondence, this makes sense only for a given cardinality. A set of axioms is called a *theory* and a theory is said to be *categorical* for a given cardinality if all models of this theory of this cardinality are isomorphic. Taking in account these details and using the related terminology we can say that axiomatizing the natural numbers means, from a model-theoretical point of view, finding a categorical theory for $\mathcal{N}$ (Categorical relatively to denumerability, since the natural numbers are typically denumerable). We will talk of *MTC-axiomatization* (MTC being an abbreviation for "model-theoretical categorical").

Although model-theoretical axiomatization was already quite clear with the work of David Hilbert in 1899 on axiomatization of geometry (Hilbert 1899) and the concept of categoricity was introduced in 1904 by Oswald Veblen (Veblen 1904), MTC-axiomatization was made perfectly precise only with the work of Tarski in model theory in the 1950s, developing in a systematic way the relation between a theory and its models (Tarski 1954a,b, 1955).

If we have a theory such that $\mathcal{N}$ is a model of this theory, but there is also a quite different structure model of this theory, then we will not say that this theory is a MTC-axiomatization of $\mathcal{N}$. In 1934 Thoralf Skolem showed that the basic axiomatization for natural numbers, *PA* (*Peano Arithmetic*), has some non-standard models, in which there are non-standard numbers coming after all the usual natural numbers (Skolem 1934). According to that, *PA* does not properly MTC-axiomatized the

---

[1] The notion of structure was promoted as the central notion of mathematics by Bourbaki (cf. Chapter 4 of the book *Théorie des ensembles* 1970, entitled *Structures*). See also Bourbaki (1948) and Corry (2004a). And Bourbaki stressed that the structure of natural numbers is not at all the simplest structure, it is a mix/combination of different structures (*carrefour de structures* in French). See our recent paper (Beziau 2017b), pointing the many aspects of the number 1 according to different structures it is merged in.

natural numbers. This result is an application of the compactness theorem, a central theorem of first-order logic.

Axiomatization in modern logic goes hand to hand with *formalization* in the sense that the axioms are formulated in a precise language having some precise properties. Axioms and other statements are expressed through *formulas*. The language of first-order logic is the main language of modern logic. There are quantifiers, connectives, relations and functions between/over objects. But in first-order logic there are no relations or functions between/over sets of objects. And in first-order logic it is not allowed to quantify over relations or functions. Basic topological concepts are typically not first-order.

In second-order logic it is possible to do so and by doing that to have a MTC-axiomatization of natural numbers. The reason why it is often not considered as satisfactory has to do with mechanization of reasoning, a concept that has been studied systematically in recursion theory. Recursion theory gives a precise definition of computability, corresponding to the informal notion of "algorithm". Second-order logic is strongly not mechanizable in particular due to the fact that the compactness theorem, according to which if a formula is a consequence of a theory it is a consequence of a finite subtheory of this theory, is not valid.

The notion of consequence can be understood in two different ways: proof-theoretically (symbolized by "$\vdash$") or model-theoretically (symbolized by "$\vDash$"). "$T \vDash F$" is understood as "All models of T are models of F" and $T \nvDash F$ as "There is a model of T which is not a model of F".[2] When we have a structure which is a model of F, we say that F is true in this structure. If we have a structure which is not a model of F, we say that F is false in this structure. By definition of classical negation (symbolized by "$\neg$"), it is equivalent as saying that $\neg F$ is true in this structure.

A theory is said to be *incomplete* when we have a formula F such that neither this formula nor its negation are a consequence of T. This formula is said to be *independent*. These notions make sense both from a proof-theoretical and model-theoretical perspectives. From the model-theoretical point of view, this means that there is a model of T in which F is true and a model of T in which F is false. So if we have a theory which is incomplete, this theory cannot be considered as a proper MTC-axiomatization of a given structure, because it has two models which essentially differ as shown by the independent formula, true in one model and false in another model.

On the other hand we may have a theory in first-order logic having different models, but complete, because the difference between models cannot necessarily be expressed in first-order logic. The incompleteness of *first-order arithmetic* (the theory formulated in first-order logic to axiomatize the structure of natural numbers $\mathcal{N}$) cannot be deduced from Skolem's theorem about non-standard models. But the fact that $\mathcal{N}$ is not MTC-axiomatizable in first-order logic can be deduced from Gödel's

---

[2]This definition was given in Tarski (1936), although at this time the notion of model was not yet completely clear. Tarski was also not yet using the symbol "$\vDash$".

(1931) proof-theoretical incompleteness theorem of arithmetic, via the completeness theorem establishing a correspondence between proof-theory and model-theory, a theorem also proved by Gödel (1930).

A structure which is MTC-axiomatizable is complete. And in first-order logic a complete theory is *decidable*, in the sense that we have an algorithm to know if a formula is a consequence or not of this theory. But a theory can also be incomplete and decidable; this is for example the case of the theory of dense order as proved by Robert Vaught in 1954. In classical propositional logic, the empty theory is incomplete and decidable: we can use truth-tables to check if a formula is a tautology or not and there are formulas, such that atomic propositions, which are independent.

Axiomatization was traditionally conceived from a *proof-theoretical* point of view, in the sense that we can *prove* all truths about a given field from these axioms. Proving meaning here a step-by-step deduction where every step is clearly explained and justified. This is how axiomatization appears in the book by Pascal (1657), who was the first to clearly describe and analyze the procedure.[3] Incompleteness can be seen as a serious drawback for proof-theoretical axiomatization. If we have an incomplete theory T for the natural numbers, there is an independent formula F. This formula expressing a statement about natural numbers is true or false. If it is true, we would like it to be a proof-theoretical consequence of T, and if it is false we would like its negation $\neg F$ to be a proof-theoretical consequence of T. But it does not work. Incompleteness can be seen here as a discrepancy between truth and proof as argued by Tarski (1969) in his famous paper "Truth and Proof".

Summarizing: from a model-theoretical point of view, the fact that it is not possible to find a first-order complete theory for the structure of the natural numbers $\mathcal{N}$ means that we cannot find some axioms formulated in first-order language that precisely catch $\mathcal{N}$; from a proof-theoretical point of view it means that we cannot prove all the truths about natural numbers from a first-order set of axioms.[4]

On the other hand the structure of natural numbers $\mathcal{N}$ is MTC-axiomatizable in second-order logic. This means that we have a clear understanding of $\mathcal{N}$, caught by a few propositions expressed in a precise language, the language of second-order logic, despite the fact that our reasoning about $\mathcal{N}$ is not mechanizable, in the sense that it cannot be fully described by a recursively enumerable system like first-order logic.

Gödel himself made the following comments: "In 1678 Leibniz made a claim of the universal characteristic. In essence it does not exist: any systematic procedure for solving problems of all kinds must be nonmechanical." (Wang 1997, p. 6.3.16); "My incompleteness theorem makes it likely that mind is not mechanical" (Wang 1997, p. 6.1.9).

---

[3] Tarski was much influenced by Blaise Pascal, in particular when writing "Sur la méthode déductive" (1937).

[4] More exactly: from a recursive set of axioms. This means we should be able to identify these axioms, we have to exclude the case where we have any infinite set of axioms, like the extreme case of all formulas true in $\mathcal{N}$, which trivially is a complete theory for $\mathcal{N}$.

The fact that our mind is not mechanical is rather good news, we are not reducible to computers. Axiomatizability in higher logic shows that we can understand things which are beyond computability. But can we understand everything, can we axiomatize everything?

8.3   Axiomatizing the Notion of Order

The relation of order between natural numbers is a *discrete* total order with first element and without last element. Discrete means that between a natural number and its immediate successor, say between 7 and 8, there is no other natural numbers. Another example of order is the strict order among rational numbers. It is radically different in the sense that between two rational numbers there is always another one, it is called a linear *dense* order. This order obeys the following axioms:

$$\forall x \quad \neg(x < x) \hspace{5cm} \text{irreflexivity}$$
$$\forall x \forall y \forall z \quad (x < y \land y < z) \to x < z \hspace{2.5cm} \text{transitivity}$$
$$\forall x \forall y \quad (x \neq y \land x < y) \to \neg(y < x) \hspace{2cm} \text{antisymmetry}$$
$$\neg \exists x \forall y \quad (x < y) \hspace{4cm} \text{no first element}$$
$$\neg \exists x \forall y \quad (y < x) \hspace{4cm} \text{no last element}$$
$$\forall x \forall y \quad (x < y \to \exists z(x < z \land z < y)) \hspace{2cm} \text{density}$$
$$\forall x \forall y \quad x \neq y \to (x < y \lor y < x) \hspace{1.5cm} \text{totality or linearity}$$

It has been shown (result originally due to Cantor) that these axioms have only one denumerable model (up to isomorphism).[5] For this reason we can say that the notion of dense order without end points is MTC-axiomatizable (in first-order logic).

From these axioms we can extract two axioms, for which we will use the sign "R" rather than "$<$":

$$\forall x \forall y \forall z \quad (xRy \land yRz) \to xRz \hspace{2cm} \text{transitivity}$$
$$\forall x \forall y \quad (x \neq y \land xRy) \to \neg yRx \hspace{2cm} \text{antisymmetry}$$

The symbol "$<$" leads us to think of irreflexive (or strict) order by contrast to the symbol "$\leq$". If we don't have the axiom $\forall x \neg(x < x)$ of irreflexivity we need a notation leaving space for our imagination, leaving the door open to various *interpretations* (a canonical concept of model theory). A relation obeying the two above axioms is called a *relation of order*.[6] Can we say that this theory, the conjunction of these two axioms, axiomatizes the notion of order?

---

[5]The axiomatization presented here is not independent in the sense that for example the axiom of antisymmetry is a consequence of the axioms of irreflexivity and transitivity.
[6]Sometimes a relation of order is defined as a relation also being reflexive. This is not a very good choice, because then the notion of strict order is contradictory.

This theory has many different models. The order can be dense or not, can have a first element or not, can be partial or not. Is the fact that we have many different models of this theory a problem for talking about axiomatization? Not necessarily. We have caught something common to all these relations of order, avoiding things opposite to this notion such as cycles. We can talk about *MT-axiomatization*, removing the question of categoricity.

MT-axiomatization is important if we want to generalize axiomatization to non-mathematical notions, for example the notion of animal. There are many "non-isomorphic" animals, different in shapes, internal features, behaviors, nevertheless one may look for axioms characterizing the very nature of the notion of animal, if any. For the relation of order we have two axioms expressed by two first-order formulas. If we put a conjunction between the two, this can be reduced to only one formula:

$$\forall x \forall y \forall z (xRy \wedge yRz) \rightarrow xRz \wedge \forall x \forall y (x \neq y \wedge xRy) \rightarrow \neg yRx$$

This is rather artificial. These two axioms correspond to two different ideas that are put together with a conjunction. This is not the same as one single axiom corresponding to a single idea from which two or more axioms can be deduced, a synthetic axiom.

In set theory there is the *axiom of abstraction* saying that any property/formula determines a set:

$$\exists x \forall y \quad (y \in x \leftrightarrow Fy)$$

The problem is that not only other basic intuitive axioms about sets are consequence of this axiom, but also all formulas, in another words: this axiom is trivial, is inconsistent. This is the famous Russell's paradox.

Reduction to a single axiom, even if it is not trivial, may appear as rather meaningless, a formal artificial game, at best showing capacity of high intellectual gymnastic. For example, Tarski (1938) provided the following single axiom for abelian group, using division as the unique primitive relation:

$$x/(y/(z/(x/y))) = z$$

This reduction is a kind of reductionism by opposition to the usual set of axioms where each axiom has a clear and definite meaning.

A more consistent and meaningful example is from the field of logic: all the properties of classical negation can be put in only one intuitive axiom, the strong reduction to the absurd. This axiom can then be decomposed in various axioms, each having a distinct meaning like elimination of double negation, *ex-falso sequitur quod libet*, etc. (for details about that, see Beziau 1994).

It is also important to stress that one axiom may have many different equivalent formulations, the most famous example being the axiom of choice (cf. Rubin and

Rubin 1963, 1985). Although all these formulations are equivalent, they correspond to different ideas. What all these ideas have in common is not clear, we cannot really say that there is one unique idea beyond/behind all these formulations. At best we can say that one formulation is more typical, more representative. This variation of formulations, of meanings, can also manifest not for only one axiom, but for a set of axioms. A notable example is the case of Boolean algebra, which can be seen either as a distributive complemented lattice or as an idempotent ring. This was discovered by Marshall Stone (1935) who was amazed by the coincidence of these two perspectives, these two different axiomatizations of the same structure.

These considerations about MT-axiomatization are important for answering the question about a single axiom for everything. This singleness can be seen as a weird singularity! If we axiomatize time using the notion of order, it is not possible to characterize it with only one axiom, considering that antisymmetry and transitivity are two different ideas that can hardly be subsumed by a unique third idea expressed in one axiom. And even if we can find one single axiom corresponding to one single idea, this axiom can be considered as equivalent to another axiom corresponding to another idea, as shown by the case of the axiom of choice. We may have different equivalent perspectives on the same reality which itself is beyond a unique particular characterization.

8.4   Axiomatizing Identity

The notion of identity seems obvious but it can be understood in different ways. Let us here consider identity as a relation between things (objectual identity). Two things can be more or less identical depending for example of how many properties they share. A definition of objectual identity, attributed to Leibniz, is that if two things have the same properties they are identical. There is a more radical notion of identity, that we will call *trivial identity*, represented by the diagram in Fig. 8.1. The relation of identity is here relating each object to itself and that's it! According to the above diagram the relation of identity does not hold between different objects. The problem is that this relation of trivial identity cannot be model-theoretically axiomatized in first-order logic (see e.g. Hodges 1983).

But there is another problem, despite the fact that we can visually represent this notion in a diagram, it is not clear that we can directly phrase it. We can say:

Every object is identical to itself and different from the others.

But this means:

Every object is identical to itself and not identical to non-identical objects.

The second part of the proposition is a tautology and since the conjunction of a proposition p with a tautology is equivalent to the proposition p, this formulation of the axiom of identity is equivalent to:

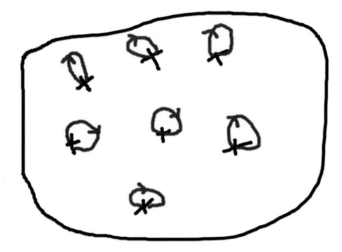

Fig. 8.1 Illustration of the identity relation

Every object is identical to itself.

And such an axiom is nothing else than the axiom of reflexivity and does not exclude reflexive relations which are not trivial identity, where two different elements can be in relation. From this perspective we can say that the relation of trivial identity corresponds to a certain situation, which we can understand through a picture but that we cannot directly phrase (for more details see Beziau 2015b).

However it is possible to formulate it in an indirect way, which can be expressed in second-order logic, MTC-axiomatizing it, saying that it is the least reflexive relation (see Manzano 1996).

This example shows different levels of understanding: at one level we pictorially understand something but cannot phrase it. At this level we cannot therefore properly talk about axiomatization. Understanding does not necessarily reduce to axiomatization. In the case of identity we are lucky that at a second level we can axiomatize it, but there can be phenomena not axiomatizable at any level, and which cannot be phrased.

8.5   Axiomatizing Causality

Leibniz has promoted the famous dictum *Nihil est sine ratione*, called the "principle of sufficient reason", which often is interpreted as *Nothing is without a cause* and that we can positively state as *Everything has a cause.*

Is it true that everything has a cause? To answer this question we must investigate what causality is. There are different ways to do that. One of them is to try to axiomatize causality. Is it possible to axiomatize causality? To do so, like for any axiomatization, we have to choose a conceptual framework.[7] One pretty natural option is to consider that causality is a binary relation between events, we can write "$a \hookrightarrow b$", read as "a causes b" or "b is caused by a".

Axiomatizing causality then means describing some basic properties of this binary relation. There are different options that we will not discuss here, just presenting one option among others (for more details see Beziau 2015a). Our objective is not to defend one particular vision of causality but to show that there are intrinsic problems which are quite independent of a particular choice.

We consider the properties described by the following formulas:

$$\forall x \quad \neg(x \hookrightarrow x) \hspace{4cm} \text{irreflexivity} \hspace{1cm} (8.1)$$

$$\forall x \forall y \quad (x \neq y \wedge x \hookrightarrow y) \rightarrow \neg(y \hookrightarrow x) \hspace{1cm} \text{antisymmetry} \hspace{1cm} (8.2)$$

$$\exists x \exists y \exists z \quad (x \hookrightarrow y) \wedge (y \hookrightarrow z) \wedge \neg(x \hookrightarrow z) \hspace{1cm} \text{non-transitivity} \hspace{1cm} (8.3)$$

$$\forall x \exists y \quad (y \hookrightarrow x) \hspace{3cm} \text{everything has a cause} \hspace{1cm} (8.4)$$

The main problem is that we have some models obeying these axioms where the relation can be seen as something radically different from causality, like the binary relation of immediate succession among the integers, i.e. $a R b$ iff $b = a + 1$.

That is another central point of the question of axiomatization: even if we do not limit axiomatization to MTC-axiomatization, extending it to MT-axiomatization, we don't want to catch things that are of a nature very different from what we have in mind. It is not clear in this case how to avoid that, if we can do that by adding further axioms.

There is another problem in this axiomatization of causality. It is with axiom (8.3), the axiom of non-transitivity, which is existential/negative. Generally when axiomatizing something we are looking for universal features.[8]

As we were saying, the above theory is just a possible axiomatization of causality. One may want for example to put an axiom saying that there is a first cause, but, in any case, whatever axioms we choose, it seems the same problem will repeat: we will have some models very different in nature of what we want to axiomatize.

Axiomatizing causality can be seen as a way to axiomatize everything. *Nihil est sine ratione* is indeed a very general principle that can be viewed as a key for the understanding of reality. But if we formalize it as axiom (8.4) above, it does not really make sense by itself, we need to add other axioms. And even adding further axioms, it is not clear at all that we are succeeding to axiomatize causality.

Rougier (1920) in his book *Les paralogismes du rationalism (Paralogisms of rationalism)* strongly criticizes such kind of general principles considering them as

---

[7]David Hilbert (1918) in his famous paper on axiomatic thought is using the expression "conceptual framework" (in German: *Fachwerk von Begriffen*), but he identifies it to the notion of theory. We prefer to use the word "theory" as referring to a set of statements that in particular can be considered as axioms.

[8]Rolando Chuaqui and Patrick Suppes (1995) have shown that classical mechanics can be axiomatized by formulas with only universal quantifiers.

meaningless and Leibniz is one of his favorite scapegoats when criticizing rationalism.[9] One of his criticisms is about the principle "The whole is bigger than the part", explaining how modern set theory resolves Galileo's paradox by making the distinction between inclusion and one-to-one correspondence. However we can say that in this case axiomatization (axiomatic set theory) helps to clarify our conceptualization.

8.6   Axiomatizing Reality

One of the difficulties for axiomatizing reality is that is has many aspects. This is what we can see looking around us here on earth, not even travelling through the universe. We may want to reduce reality to few objects, few phenomena, but that's not so easy without a heavy reductionism, like physicalism.

We can axiomatize physics, Hilbert himself did a lot of work in that direction (see Corry 2004b). But to axiomatize the reducibility, let's say of biological phenomena, not to say psychological phenomena, to physical phenomena is not that easy. We can argue that physicalism will be seriously supported only when such kind of axiomatization is provided.

Einstein is reported to have claimed: "The grand aim of all science is to cover the greatest number of empirical facts by logical deduction from the smallest possible number of hypotheses or axioms" (Barnett 1948). The axiomatization mentioned here is not what we have called model-theoretical axiomatization, it is rather proof-theoretical axiomatization.

However the theory of general relativity can be model-theoretically axiomatized in various ways in first-order logic (see Andréka, Madarsz and Németi 2007). As it is known, Gödel (1949) has shown that Einstein general theory of relativity has some non-standard models, so it is not a MTC-axiomatization. It does not fully grasp reality because it admits different incompatible models. To solve the problem it is necessary to reduce the number of models. If it possible to do that in a natural way, by finding a categorical axiomatization of the universe based on some intuitive axioms, not by artificially adding an *ad hoc* axiom eliminating rotating universe, is an open question. And someone who believes that our universe is really rotating should find an axiomatization which does not admit a standard model according to which we cannot come back in the past.

Axiomatizing the (physical) universe is something, axiomatizing (biological) life is another thing. Physics can in some sense be seen as something merely geometrical, part of mathematics. It is easier to fix a conceptual framework as the basis for an axiomatization of physics, than of biology. Axiomatization of biology is still something very much experimental, although it was initiated at the beginning of the

---

[9]Rougier (1889–1992) was a good friend of Moritz Schlick and one of the main promoters of the Vienna Circle. He organized in 1935 at the Sorbonne in Paris a big congress on scientific philosophy with the participation of Schlick, Carnap, Neurath, Russell, Tarski, Lindenbaum, etc. The result was a 8-volume book: *Actes du Congrès International de Philosophie Scientifique – Sorbonne, Paris, 1935.*

20th century. Tarski first love was neither mathematics, nor logic, but biology and since he was found of the axiomatic method he encouraged people to axiomatize biology in particular Joseph Henri Woodger (1937).

Someone may want to develop a general theory/axiomatization of everything using a very abstract theory not directly ontologically committed with the nature of its objects, for example set theory. But even in the case of someone convincingly succeeding to argue that everything in reality is (or can be interpreted as) a set, does this mean that set theory can axiomatize reality? There is an ambiguity here. If we just stay at the mathematical level, we can say that the notion of a group can be conceived and defined using set theory, but it does not make properly sense to say that an axiomatic theory of set, let's say *ZF*, axiomatizes group theory, in particular the axioms of group theory are not a consequence of the axioms of *ZF* (for more details see Beziau 2002).

8.7   Axiomatizing Reasoning

Human beings have been characterized as rational animals, logical animals, animals able to reason. Reasoning is a basic feature of human beings and reasoning can be considered as the backbone of thought. "Logic is the anatomy of thought" is a dictum attributed to Locke, logic has been named the *Art of Thinking* (cf. Arnauld and Nicole 1662) and Boole, considered as one of the main originators of modern logic, wrote a famous book called *The Laws of Thought* (1854) (see Beziau 2010, 2017a).

From this perspective axiomatizing reasoning may be interpreted as axiomatizing human beings and also as indirectly axiomatizing reality, considering that reasoning is the basic tool to capture, describe, understand reality.

But can we axiomatize reasoning? Can we find some axioms describing the reality of reasoning, the way we are reasoning? To answer this question we will here again consider axiomatization from a model-theoretical point of view. Can we find some axioms, whether in first-order logic or in second-order logic, which are a MTC-axiomatization of reasoning?

As for other fields, we have to set up a conceptual framework. We consider the one directly inspired by Tarski's theory of consequence operator initiated in Tarski (1928), where we have an abstract consequence relation, that we will express using the notation "$T \Vdash F$". Proof-theoretical consequence (symbolized by "$\vdash$") and model-theoretical consequence (symbolized by "$\vDash$") can both be seen as particular models of this abstract consequence relation. By modeling model-theoretical consequence we are at a meta-level. We can also say that we are axiomatizing axiomatization (see Beziau 2006).

We may fix some general axioms for this abstract consequence relation, such as the three following ones (due to Tarski):

$T \Vdash F$ when F is an element of T reflexivity

If $T \Vdash F$ and T is included in U, then $U \Vdash F$ monotonicity

If $T \Vdash F$ for every F in U and $U \Vdash G$, then $T \Vdash G$ transitivity

These axioms make sense if we have in view a proof-theoretical notion of consequence defined using what is called an "Hilbert system". They also hold for a model-theoretical notion of consequence defined using the basic idea of model inclusion (cf. Tarski 1936). So these axioms are very general. They hold not only independently of a specific logical language corresponding to a given conceptual framework, but also independently of the way the consequence relation is originally conceived.

These axioms admit many different models, so they are not going on the direction of a MTC-axiomatization of reasoning. This can be done by adding further axioms. For example with the two following axioms

$$T \Vdash p \rightarrow q \; iff \; T, p \Vdash q$$

If $T, \neg p \Vdash q$ and $T, \neg p \Vdash \neg q$ then $T \Vdash p$

we succeed to fully axiomatize classical propositional logic. This does not mean that we have axiomatized reasoning, unless we believe that reasoning reduces to classical propositional logic. Hard to believe! At best one may believe that reasoning reduces to the one described by first-order logic and we can add further axioms about quantifiers to axiomatize this system.

But many other different logical systems have been developed in the last 100 years and it is not clear what the correct one is, if any. Anyway, a general theory of abstract consequence relation allows us to master all these systems, by MT-axiomatizing them.

However there may be doubts about the validity of the three Tarski's axioms. In particular the second one does not hold for logical systems which have been qualified as "non-monotonic" for this very reason. Shall we delete this axiom? And then do not we have an axiomatization which is too general, not characterizing reasoning, admitting models having nothing to do with reasoning?

It is very difficult to sustain that we can axiomatize reasoning unless we believe that there is a very specific right kind of reasoning that can be described within a precise conceptual framework.

8.8   Axiomatizing and Understanding

To conclude we can say that the axiomatic method, that we can summarize in one word, *axiomatizing*, is a key to understanding, not only by catching a structure up to isomorphism, but also by catching a notion having many different aspects, like a relation of order.

And it is important to make a clear distinction between axiomatization and mechanization of thought. Understanding certainly does not reduce to computability. Our analysis of axiomatization also clearly shows that understanding does not reduce to the singularity of the singleness of an axiom. Although a key feature of axiomatizing is to make a reduction to some few statements about few concepts, the axiomatic method is not monotheist. Moreover, there are different ways of understanding which, although equivalent, have each their own value. The axiom method is not one-sided.

Finally, there are "things", like reasoning itself, that we cannot properly axiomatize and that maybe are beyond our understanding, although understanding does not necessarily reduce to axiomatization.

Acknowledgements The origin of this paper is a talk presented at the meeting *Kurt Gödel's Legacy: Does Future lie in the Past?* which took place July 25–27, 2019 in Vienna, Austria. Thanks to the organizers of the event and also thanks to Jan Zygmunt, Robert Purdy and Jorge Petrucio Viana for useful comments and information.

References

Actes du Congrès International de Philosophie Scientifique – Sorbonne, Paris, 1935 (1936).

Andréka, H., Madarsz, J., & Németi, I. (2007). Logic of space-time and relativity theory. In M. Aiello, I. Pratt-Hartmann, & J. van Benthem (Eds.), *Handbook of spatial logics* (pp. 607–711). Heidelberg: Springer.

Arnauld, A., & Nicole, P. (1662). *La logique ou l'art de penser*. Paris: Savreux.

Barnett, L. (1948). *The Universe and Dr Einstein*. New York: William Morrow.

Beziau, J.-Y. (1994). Théorie législative de la négation pure. *Logique et Analyse, 147–148,* 209–225.

Beziau, J.-Y. (2002). La théorie des ensembles et la théorie des catégories: Présentation de deux soeurs ennemies du point de vue de leurs relations avec les fondements des mathématiques. *Boletín de la Asociación Matemática Venezolana, 9,* 45–53.

Beziau, J.-Y. (2006). Les axiomes de Tarski. In R. Pouivet & M. Rebuschi (Eds.), *La philosophie en Pologne 1918–1939* (pp. 135–149). Paris: Vrin.

Beziau, J.-Y. (2010). Logic is not logic. *Abstracta, 6,* 73–102.

Beziau, J.-Y. (2015a). Modelling causality. *Conceptual clarifications tributes to Patrick Suppes (1922–2014)* (pp. 187–205). London: College Publications.

Beziau, J.-Y. (2015b). Panorama de l'identité. *Al Mukhatabat, 14,,* 205–219.

Beziau, J.-Y. (2017a). Being aware of rational animals. In G. Dodig-Crnkovic & R. Giovagnoli (Eds.), *Representation and reality: Humans, animals and machines* (pp. 319–331). Cham: Springer.

Beziau, J.-Y. (2017b). MANY 1 – A transversal imaginative journey across the realm of mathematics. In: M. Chakraborty & M. Friend (Eds.), *Special issue on Mathematical Pluralism of the Journal of Indian Council of Philosophical Research, 34*(2), 259–287.

Boole, G. (1854). *An Investigation of the laws of thought, on which are founded the mathematical theories of logic and probabilities*. London: Macmillan & Co.

Bourbaki, N. (1948). L'architecture des mathématiques -La mathématique ou les mathématiques. In F. le Lionnais (Ed.), *Les grands courants de la pensée mathématique* (pp. 35–47). Cahier du Sud.

Bourbaki, N. (1970). *Théorie des ensembles*. Paris: Hermann.

Chuaqui, R., & Suppes, P. (1995). Free-variable axiomatic foundations of infinitesimal analysis: A fragment with finitary consistency proof. *The Journal of Symbolic Logic, 60,* 122–159.

Corry, L. (2004a). *Modern algebra and the rise of mathematical structures* (2. edn.). Basel: Birkhäuser.

Corry, L. (2004b). *David Hilbert and the axiomatization of physics (1898–1918)*. Dordrecht: Springer.

Dedekind, R. (1888). *Was sind und was sollen die Zahlen?*. Braunschweig: Vieweg.

Gödel, K. (1930). DieVollständigkeit der Axiome des logischen Funktionenkalküls. *Monatshefte für Mathematik und Physik, 37*, 349–60.

Gödel, K. (1931). Über formal unentscheidbare Sätze der Principia Mathematica und verwandter Systeme, I. *Monatshefte für Mathematik und Physik, 38*, 173–98.

Gödel, K. (1949). An example of a new type of cosmological solutions of Einstein's field equations of gravitation. *Reviews of Modern Physics, 21*, 447–450.

Gonseth, F. (1939). La méthode axiomatique. *Bulletin de la Société Mathématique de France, 67*, 43–63.

Halbach, V. (2011). *Axiomatic theories of truth*. Cambridge: Cambridge University Press.

Hilbert, D. (1899). *Grundlagen der Geometrie*. Leipzig: Teubner.

Hilbert, D. (1918). Axiomatisches Denken. *Mathematische Annalen, 78*, 405–415.

Hintikka, J. (2011). What is the axiomatic method? *Synthese, 183*, 69–85.

Hodges, W. (1983). Elementary predicate logic. In D. Gabbay & F. Guenthner (Eds.), *Handbook of Philosophical Logic I*. Dordrecht: Reidel.

Manzano, M. (1996). *Extensions of first order logic*. Cambridge: Cambridge University Press.

Pascal, B. (1657). *De l'esprit géométrique et de l'art de persuader*. Œuvres complètes, J. Chevalier (Ed.), 1962. Paris: Bibliothèque de la Plèiade.

Peano, G. (1889). *Arithmetices Principia, Novo Methodo Exposita*. Rome: Fratres Bocca.

Peirce, C. (1881). On the logic of number. *American Journal of Mathematics, 4*(1), 85–95.

Rougier, L. (1920). *Les paralogismes du rationalisme*. Paris: Félix Alcan.

Rubin, H., & Rubin, J. (1963). *Equivalents of the axiom of choice I*. Amsterdam: North Holland.

Rubin, H., & Rubin, J. (1985). *Equivalents of the axiom of choice II*. Amsterdam: North Holland.

Skolem, T. (1934). Über die Nicht-charakterisierbarkeit der Zahlenreihe mittels endlich oder abzählbar unendlich vieler Aussagen mit ausschliesslich Zahlenvariablen. *Fundamenta Mathematicae, 23*, 150–161.

Stone, M. (1935). Subsumption of the theory of Boolean algebras under the theory of rings. *Proceedings of National Academy of Sciences, 21*, 103–105.

Tarski, A. (1928). Remarques sur les notions fondamentales de la méthodologie des mathématiques. *Annales de la Société Polonaise de Mathématique, 7*, 270–272.

Tarski, A. (1936). O pojęciu wynikania logicznego. *Przegad Filozoficzny, 39*, 58–68.

Tarski, A. (1937). Sur la méthode déductive. *Travaux du IXe Congrès International de Philosophie VI* (pp. 95–103). Paris: Hermann.

Tarski, A. (1938). Ein Beitrag zur Axiomatik der Abelschen Gruppen. *Fundamenta Mathematicae, 30*, 253–256.

Tarski, A. (1944). The Semantic conception of truth and the foundations of semantics. *Philosophy and Phenomenological Research, 4*, 341–376.

Tarski, A. (1954a). Contributions to the theory of models. I. *Indagationes Mathematicae, 16*, 572–581.

Tarski, A. (1954b). Contributions to the theory of models. II. *Indagationes Mathematicae, 16*, 582–588.

Tarski, A. (1955). Contributions to the theory of models. III. *Indagationes Mathematicae, 17*, 56–64.

Tarski, A. (1969). Truth and proof. *Scientific American, 220*(6), 63–77.

Vaugth, R. (1954). Applications of the Löwenheim-Skolem-Tarski theorem to problems of completeness and decidability. *Indagationes Mathematicae, 16*, 467–472.

Veblen, O. (1904). A system of axioms for geometry. *Transactions of the American Mathematical Society, 5*, 343–384.

Wang, H. (1997). *A logical journey – From Gödel to philosophy*. Cambridge: MIT Press.

Woodger, J. (1937). *The axiomatic method in biology, with appendices by Alfred Tarski & W.J. Floyd*. Cambridge: Cambridge University Press.

Unerklärliche Wahrheiten

9

Marco Hausmann

9.1 Einleitung

9.1.1 Erwartungen an die Wissenschaft

Manche Menschen haben hohe Erwartungen an die Wissenschaft. Manche Menschen erwarten von der Wissenschaft, dass sie die Welt *vollständig beschreiben* wird. Die Wissenschaft wird diese Erwartung aber sicher nie erfüllen – zumindest dann nicht, wenn sie versucht, alle Wahrheiten über die Welt *einzeln aufzuzählen*. Denn es gibt überabzählbar viele Wahrheiten (weil es mindestens genauso viele Wahrheiten wie reelle Zahlen gibt[1] und weil es, wie Cantor (1892) gezeigt hat, überabzählbar viele reelle Zahlen gibt). Die Wissenschaft wird diese Erwartung also, wenn überhaupt, nur dann erfüllen, wenn sie versucht, endlich viele Wahrheiten aufzuzählen, aus denen man dann zumindest im Prinzip alle Wahrheiten ableiten kann. Man kann die Welt also nicht vollständig beschreiben, wenn man alle Wahrheiten einzeln aufzählt (denn dann würde man nie an ein Ende kommen); man kann die Welt, wenn überhaupt, nur vollständig beschreiben, wenn man endlich viele Wahrheiten (Axiome, Gesetze, …) aufzählt, aus denen man dann zumindest im Prinzip alle Wahrheiten ableiten kann.

---

[1]Es gibt mindestens genauso viele Wahrheiten wie reelle Zahlen, denn, um nur ein Beispiel zu nennen, jeder reellen Zahl entspricht entweder die Wahrheit, dass diese reelle Zahl eine Primzahl ist, oder aber die Wahrheit, dass diese reelle Zahl keine Primzahl ist.

Vielen Dank an Simon Voderholzer und Lukas Hallmann für die wertvolle Hilfe bei der Beschaffung der Literatur und vielen Dank an Geert Keil für die hilfreiche Diskussion zum Aufsatz (auch wenn es für eine inhaltliche Neuausrichtung des Aufsatzes leider schon zu spät war).

M. Hausmann (✉)
Ludwig-Maximilians-Universität München, München, Deutschland
E-mail: marco.hausmann1@gmx.de

© Der/die Autor(en), exklusiv lizenziert durch Springer-Verlag GmbH, DE, ein Teil von Springer Nature 2021
O. Passon und C. Benzmüller (Hrsg.), *Wider den Reduktionismus,*
https://doi.org/10.1007/978-3-662-63187-4_9

Kurt Gödel hat allerdings gezeigt, dass diese Erwartung an die Wissenschaft zu hoch ist. Die Wissenschaft wird die Welt *niemals vollständig beschreiben,* indem sie einzelne Wahrheiten (Axiome, Gesetze, …) aufzählt, aus denen man dann alle Wahrheiten ableiten kann. Denn um die Welt vollständig zu beschreiben, müsste die Wissenschaft zumindest die Axiome der Arithmetik aufzählen. Gödel (1931) hat aber gezeigt, dass man aus den Axiomen der Arithmetik nicht ableiten kann, dass die Axiome der Arithmetik konsistent sind, wenn die Axiome der Arithmetik konsistent sind.[2] Nun sind die Axiome der Arithmetik aber konsistent. Man wird aus den endlich vielen Wahrheiten (Axiomen, Gesetzen, …), die die Wissenschaft aufzählen muss, um die Welt vollständig zu beschreiben, also niemals *alle* Wahrheiten ableiten können. Denn es gibt dann, wie Gödel gezeigt hat, zumindest eine Wahrheit, die man aus diesen einzelnen Wahrheiten (Axiomen, Gesetzen, …) niemals ableiten können wird – die Wahrheit, dass die Axiome der Arithmetik konsistent sind.

Manche Menschen haben aber nach wie vor hohe Erwartungen an die Wissenschaft. Manche Menschen erwarten von der Wissenschaft, dass sie die Welt *vollständig erklären* wird. Die Wissenschaft wird diese Erwartung aber sicher nie erfüllen – zumindest dann nicht, wenn sie versucht, alle Wahrheiten über die Welt *einzeln aufzuzählen und zu erklären* (denn es gibt, wie wir bereits gesehen haben, überabzählbar viele Wahrheiten). Dennoch könnte die Wissenschaft versuchen, dieser Erwartung dadurch gerecht zu werden, dass sie endlich viele Wahrheiten aufzählt, mit denen man dann zumindest im Prinzip alle Wahrheiten erklären kann. Ziel meiner Überlegungen ist es, zu zeigen, dass dieser Versuch aus prinzipiellen Gründen zum Scheitern verurteilt ist. Wer von der Wissenschaft erwartet, dass sie die Welt *vollständig erklären* wird, der hat eine zu hohe Erwartung an die Wipricht manssenschaft. Die Wissenschaft wird diese Erwartung aus prinzipiellen Gründen nie erfüllen.

9.1.2   Die These des Reduktionismus

Eine heutzutage weit verbreitete These ist die These des *Reduktionismus.* Der Reduktionismus behauptet allgemein genommen, dass alle Wahrheiten mithilfe von Wahrheiten über einen bestimmten Bereich erklärt werden können.[3] Es gibt viele verschiedene Versionen des Reduktionismus. Eine heutzutage weit verbreitete Version des Reduktionismus ist die These des *Naturalismus.* Die These des Naturalismus ist nach meinem Verständnis die These, dass alle Wahrheiten mithilfe von Wahrheiten über natürliche Phänomene erklärt werden können. Eine heutzutage ebenso weit verbreitete (und oft mit dem Naturalismus in Verbindung gebrachte) Version des Reduktionismus ist die These des Physikalismus. Die These des Physikalismus ist

---

[2]Für unsere Zwecke genügt diese grobe Formulierung des zweiten Unvollständigkeitssatzes. Für eine genauere Darstellung vgl. z. B. Boolos (1996).

[3]In diesem Zusammenhang spricht man davon, dass man alle Wahrheiten mithilfe von Wahrheiten über einen bestimmten Bereich *reduktiv erklären* kann bzw. dass man alle Wahrheiten auf Wahrheiten aus einem bestimmten Bereich *reduzieren* kann.

nach meinem Verständnis die These, dass alle Wahrheiten mithilfe von Wahrheiten über physikalische Phänomene erklärt werden können. Eine heutzutage zwar selten noch mit diesem Namen in Verbindung gebrachte, aber dennoch weiterhin einflussreiche Version des Reduktionismus ist die These des Atomismus. Die These des Atomismus ist nach meinem Verständnis die These, dass alle Wahrheiten mithilfe von Wahrheiten über die kleinsten physikalischen Teilchen erklärt werden können.

Die These des Reduktionismus würde ohne Zweifel, wenn sie wahr wäre, die Hoffnung wecken, dass die Wissenschaft eines Tages die Welt vollständig erklären können wird. Denn wenn die These des Reduktionismus wahr wäre, dann gäbe es einzelne Wahrheiten (zum Beispiel Wahrheiten über natürliche Phänomene, Wahrheiten über physikalische Phänomene, oder Wahrheiten über die kleinsten physikalischen Teilchen), mit denen man *alle* Wahrheiten erklären kann. Es würde dann genügen, diese Wahrheiten aufzuzählen, um die Welt vollständig erklären zu können.[4] Der Wissenschaft wären dann in ihrem Versuch, die Welt vollständig zu erklären, keine prinzipiellen Grenzen gesetzt.

Ziel meiner Überlegungen ist es, zu zeigen, dass die These des Reduktionismus falsch ist, weil sie auf einer falschen Voraussetzung beruht. Die These des Reduktionismus beruht auf der Voraussetzung, dass alle Wahrheiten erklärt werden können. Ich werde dagegen versuchen, zu zeigen, dass es Wahrheiten gibt, die nicht erklärt werden können. Die These des Reduktionismus ist nach meiner Überzeugung also falsch.

Meine Überlegungen sind in drei Abschnitte gegliedert. In Abschn. 9.2 werde ich, um Missverständnisse aus dem Weg zu räumen, mein Verständnis von der These des Reduktionismus kurz erläutern. In Abschn. 9.3 werde ich in Anlehnung an einen Gedankengang, der auf Alonzo Church zurückgeht, ein Argument gegen den Reduktionismus entwickeln und zeigen, dass man mithilfe modifizierter Versionen dieses Arguments auch gegen schwächere Versionen des Reduktionismus argumentieren kann. In Abschn. 9.4 werde ich erläutern, welche Auswirkungen meine Überlegungen für unsere Erwartungen an die Wissenschaft haben. Ich werde dafür argumentieren, dass der Versuch, die Welt mithilfe der Wissenschaft vollständig zu erklären, aus prinzipiellen Gründen zum Scheitern verurteilt ist. Nach einem knappen Fazit in Abschn. 9.5 wird in Abschn. 9.6 eine formallogische Ausarbeitung meiner wichtigsten Argumente gegeben.

---

[4]Wenn man diese Wahrheiten *nicht* aufzählen könnte, zum Beispiel weil es überabzählbar viele Wahrheiten wären, dann wäre der Traum von der vollständigen Erklärung der Welt durch die Wissenschaft natürlich trotzdem geplatzt. Die These des Reduktionismus impliziert nach meinem Verständnis, dass es Wahrheiten über einen bestimmten Bereich gibt, mit denen man alle Wahrheiten erklären kann, die These des Reduktionismus impliziert nach meinem Verständnis aber *nicht*, dass es *endlich viele* Wahrheiten über einen bestimmten Bereich gibt, mit denen man alle Wahrheiten erklären kann.

9.2   Kurze Erläuterung zur These des Reduktionismus

Klar ist, dass die These des Reduktionismus unterschiedlich aufgefasst wird (und
zwar unabhängig davon, ob von einem naturalistischen, physikalistischen oder ato-
mistischen Reduktionismus die Rede ist). Es ist deshalb nötig, mein Verständnis von
der These des Reduktionismus kurz zu erläutern. Hierzu drei kurze Bemerkungen:

1. Die These des Reduktionismus ist nicht als eine spekulative Vorhersage über die
   Zukunft, sondern als Aussage über eine prinzipielle Möglichkeit aufzufassen.
   Der Reduktionist behauptet nicht, dass wir Menschen *irgendwann einmal* für
   alle Wahrheiten eine Erklärung aus einem bestimmten Bereich *geben werden*
   (denn hierfür müssten wir Menschen, wie gesagt, überabzählbar viele Wahrheiten
   aufzählen), sondern er behauptet lediglich, dass wir Menschen für alle Wahrheiten
   im Prinzip eine Erklärung aus einem bestimmten Bereich *geben könnten*.[5]
2. Die These des Reduktionismus ist nicht die These, dass wir Menschen für alle
   Wahrheiten eine womöglich falsche Erklärung aus einem bestimmten Bereich
   geben können, sondern die These, dass wir Menschen für alle Wahrheiten eine
   korrekte Erklärung aus einem bestimmten Bereich geben können. Denn man kann
   zwar Verbrennungsvorgänge mithilfe von Phlogiston, oder Bahnabweichungen
   des Planeten Merkur mithilfe des Planeten Vulkan erklären, man kann Verbrennungsvorgänge aber nicht korrekt mithilfe von Phlogiston erklären, und man kann
   Bahnabweichungen des Planeten Merkur nicht korrekt mithilfe des Planeten Vulkan erklären.
   Die These des Reduktionismus ist also die These, dass *wir Menschen* für alle
   Wahrheiten mithilfe von Wahrheiten aus einem bestimmten Bereich *im Prinzip*
   eine korrekte Erklärung *geben könnten.*
3. Nicht nur im Alltag, auch in den Wissenschaften haben wir es für gewöhnlich mit
   partiellen (und nicht mit *vollständigen*) Erklärungen zu tun.[6] Die Erklärung, dass
   die Erde sich um die Sonne dreht, weil es Krümmungen in der Raumzeit gibt,
   erklärt nur zum Teil (und nicht vollständig), warum sich die Erde um die Sonne
   dreht (denn um vollständig zu erklären, warum sich die Erde um die Sonne dreht,
   müsste man zusätzlich anführen, dass die Raumzeit *um die Sonne* gekrümmt ist,
   etc.). Es ist klar, dass der Reduktionist nicht nur behaupten will, dass alle Wahrheiten mithilfe von Wahrheiten über einen bestimmten Bereich *partiell* erklärt
   werden können. Denn wenn alle Wahrheiten mithilfe von Wahrheiten über einen
   bestimmten Bereich *nur* partiell (aber *nicht vollständig*) erklärt werden könnten,
   dann wäre es falsch, zu behaupten, dass man alle Wahrheiten mithilfe von Wahr-

---

[5]Physikalische Reduktionisten würden zum Beispiel behaupten, dass wir Menschen für alle Wahrheiten im Prinzip eine physikalische Erklärung geben *könnten, wenn wir eine vollkommene physikalische Theorie hätten.* Sie würden hinzufügen, dass es *möglich* ist, eine vollkommene physikalische
Theorie zu entwickeln – auch dann, wenn wir Menschen niemals eine vollkommene physikalische
Theorie entwickeln werden.

[6]Zur Unterscheidung zwischen partiellen und vollständigen Erklärungen vgl. Schnieder (2011, S.
450 f.) und Fine (2012, S. 50).

heiten über diesen Bereich *reduktiv erklären* kann bzw. dass man alle Wahrheiten auf Wahrheiten aus diesem Bereich *reduzieren* kann. Reduktionisten wollen weitaus mehr behaupten. Reduktionisten wollen behaupten, dass alle Wahrheiten mithilfe von Wahrheiten über einen bestimmten Bereich *vollständig* erklärt werden können. Die These des Reduktionismus ist also die These, dass wir Menschen für alle Wahrheiten mithilfe von Wahrheiten aus einem bestimmten Bereich *im Prinzip* eine *korrekte und vollständige* Erklärung geben *könnten*. Kurz gefasst:

Reduktionismus Alle Wahrheiten können mithilfe von Wahrheiten über einen bestimmten Bereich *vollständig* erklärt werden.

Ich werde dagegen dafür argumentieren, dass der Reduktionismus auf einer falschen Voraussetzung beruht – auf der Voraussetzung, dass alle Wahrheiten vollständig erklärt werden können.

9.3   Ein Argument gegen den Reduktionismus

Mein Argument gegen den Reduktionismus ist denkbar einfach: Wenn der Reduktionismus wahr ist, dann können alle Wahrheiten mithilfe von Wahrheiten über einen bestimmten Bereich vollständig erklärt werden. Wenn aber alle Wahrheiten mithilfe von Wahrheiten über einen bestimmten Bereich vollständig erklärt werden können, dann können alle Wahrheiten vollständig erklärt werden. Es gibt aber Wahrheiten, die nicht vollständig erklärt werden können. Der Reduktionismus ist also falsch.[7]

Dieses Argument gegen den Reduktionismus ist zunächst einmal völlig wertlos – es sei denn, es gibt gute Gründe dafür anzunehmen, dass es Wahrheiten gibt,

---

[7]Es ist in meinen Augen aufschlussreich, zwischen Erklärungen im uneigentlichen Sinn und Erklärungen im eigentlichen Sinn zu unterscheiden. Im uneigentlichen Sinn erklärt man genau dann, warum p, wenn man etwas behauptet, das erklärt, warum p. Im eigentlichen Sinn erklärt man dagegen genau dann, warum p, wenn man etwas behauptet, das erklärt, warum p, und *wenn man zusätzlich behauptet, dass das, was man behauptet, erklärt, warum p.* Wenn in meinen Überlegungen davon die Rede ist, dass jemand erklärt, warum p, dann ist damit gemeint, dass er oder sie im *uneigentlichen Sinn* erklärt, warum p.

In meinen Überlegungen argumentiere ich also dafür, dass es (wenigstens) eine Wahrheit gibt, die niemand im uneigentlichen Sinn erklären kann. Wichtig ist, zu beachten: Es ist notwendig, dass man das, was man im eigentlichen Sinn erklärt, auch im uneigentlichen Sinn erklärt (denn wer im eigentlichen Sinn erklärt, warum p, der behauptet etwas, das erklärt, warum p). Wenn es eine Wahrheit gäbe, die von jemandem im eigentlichen Sinn erklärt werden kann, dann gäbe es also auch eine Wahrheit, die von jemandem im uneigentlichen Sinn erklärt werden kann. In meinen Überlegungen argumentiere ich also (zumindest indirekt) auch dafür, dass es (wenigstens) eine Wahrheit gibt, die von niemandem im *eigentlichen Sinn* erklärt werden kann. Egal, ob mit der These des Reduktionismus Erklärungen im eigentlichen oder im uneigentlichen Sinn gemeint sind: Das Ergebnis meiner Überlegungen ist aus diesem Grund in jedem Fall relevant für die These des Reduktionismus.

die nicht vollständig erklärt werden können. Denn nur dann können Reduktionisten dieses Argument nicht einfach problemlos zurückweisen. Um zu zeigen, dass es Wahrheiten gibt, die nicht vollständig erklärt werden können, werde ich ein Argument aufgreifen, das auf Alonzo Church zurückgeht und das zu zeigen versucht, dass es Wahrheiten gibt, die wir nicht wissen *können* (da es Wahrheiten gibt, die wir nicht wissen). Ich werde dieses Argument aufgreifen und umformulieren, um zu zeigen, dass es Wahrheiten gibt, die nicht vollständig erklärt werden *können* (da es Wahrheiten gibt, die nicht vollständig erklärt werden). Ich werde das umformulierte Argument außerdem mehrmals umformulieren, um gegen schwächere Versionen des Reduktionismus zu argumentieren (und um dabei zu zeigen, dass es *in jedem Bereich* Wahrheiten gibt, die nicht vollständig erklärt werden können, sowie dass es Wahrheiten gibt, die *nicht einmal zum Teil* erklärt werden können).

9.3.1 Es gibt unerklärliche Wahrheiten

Alonzo Church hat in einem anonymen Gutachten (das erst kürzlich auf Church zurückgeführt werden konnte) einen Gedankengang entwickelt, der zeigt, dass es (wenigstens) eine Wahrheit gibt, die niemand wissen *kann,* wenn es eine Wahrheit gibt, die niemand weiß.[8] Denn, so der Gedankengang, wenn es eine Wahrheit gibt, die niemand weiß, dann *kann* niemand wissen, dass diese Wahrheit wahr ist und von niemandem gewusst wird. Denn wenn jemand wüsste, dass diese Wahrheit wahr ist und von niemandem gewusst wird, dann wüsste jemand, dass diese Wahrheit wahr ist, und dann wüsste jemand, dass diese Wahrheit von niemandem gewusst wird. Es wäre dann aber *sowohl wahr,* dass diese Wahrheit von jemandem gewusst wird (denn dann wüsste jemand, dass diese Wahrheit wahr ist), *als auch nicht wahr,* dass diese Wahrheit von jemandem gewusst wird (denn jemand kann nur wissen, dass diese Wahrheit von niemandem gewusst wird, wenn diese Wahrheit *tatsächlich* von niemandem gewusst wird). Es *kann* also niemand wissen, dass diese Wahrheit wahr ist und von niemandem gewusst wird. Es gibt also Wahrheiten, die niemand wissen *kann,* wenn es Wahrheiten gibt, die niemand weiß.

   Dieser Gedankengang lässt sich, nach meiner Überzeugung, aufgreifen und umformulieren, um zu zeigen, dass es Wahrheiten gibt, die niemand vollständig

---

[8]Das Gutachten, das Church verfasst hat, war an Frederic Fitch gerichtet. Fitch (1963) hat diesen Gedankengang dann aufgenommen, verallgemeinert und veröffentlicht. Für die historischen Hintergründe vgl. Salerno (2009). Zwar hat Fitch betont, dass er diesen Gedankengang einem anonymen Gutachten verdankt. Dennoch spricht man heutzutage meist vom „Fitch-Paradox". Seit bekannt ist, dass das anonyme Gutachten auf Church zurückgeht, spricht man manchmal auch vom „Church-Fitch-Paradox". Es ist allerdings, wie Williamson (2000, S. 269 f.) zeigt, etwas überzogen, in diesem Zusammenhang von einem „Paradox" zu sprechen. Aus diesem Grund ziehe ich es vor, in meinen Überlegungen nicht vom „Fitch-Paradox" oder vom „Church-Fitch-Paradox", sondern ganz einfach vom „Gedankengang, der auf Church zurückgeht" zu sprechen.

erklären *kann*.[9] Denn wenn es eine Wahrheit gibt, die niemand vollständig erklärt, dann *kann* niemand vollständig erklären, warum diese Wahrheit wahr ist und von niemandem vollständig erklärt wird. Denn wenn jemand vollständig erklären würde, warum diese Wahrheit wahr ist und von niemandem vollständig erklärt wird, dann würde jemand vollständig erklären, warum diese Wahrheit wahr ist, und dann würde jemand vollständig erklären, warum diese Wahrheit von niemandem vollständig erklärt wird. Es wäre dann aber *sowohl wahr,* dass diese Wahrheit von jemandem vollständig erklärt wird (denn dann würde jemand vollständig erklären, warum diese Wahrheit wahr ist), *als auch nicht wahr,* dass diese Wahrheit von jemandem vollständig erklärt wird (denn jemand kann nur vollständig erklären, warum diese Wahrheit von niemandem vollständig erklärt wird, wenn diese Wahrheit *tatsächlich* von niemandem vollständig erklärt wird). Es *kann* also niemand vollständig erklären, warum diese Wahrheit wahr ist und von niemandem vollständig erklärt wird. Es gibt also Wahrheiten, die niemand vollständig erklären *kann,* wenn es Wahrheiten gibt, die niemand vollständig erklärt. Nun gibt es aber Wahrheiten, die niemand vollständig erklärt.[10] Es gibt also Wahrheiten, die niemand vollständig erklären *kann*.

Eine formallogische Ausarbeitung meiner Argumentation befindet sich in Abschn. 9.6. Mein Argument beruht auf zwei zentralen Annahmen. Die erste lautet, dass man eine Konjunktion nur vollständig erklären kann, wenn man alle ihre Konjunkte vollständig erklärt:

Konjunktionsprinzip Wenn jemand vollständig erklärt, warum p & q, dann erklärt jemand vollständig, warum p, und dann erklärt jemand vollständig, warum q.

Denn mein Argument geht von einer Wahrheit aus, die niemand vollständig erklärt. Es geht dann davon aus, dass, wenn jemand vollständig erklären würde, warum

---

[9]Salerno (2018, S. 461 f.) hat kürzlich in einem kurzen Absatz vorgeschlagen (nachdem ich die Arbeit an meinem Aufsatz bereits begonnen hatte), den Gedankengang, der auf Church zurückgeht, nicht nur auf Wissen, sondern auch auf Erklärungen anzuwenden. Er hat diesen Vorschlag aber weder im Detail ausgearbeitet, noch ist er auf die Bedeutung dieses Vorschlags für die These des Reduktionismus oder für unsere Erwartungen an die Wissenschaft eingegangen. Ich unterscheide in meinem Aufsatz, anders als Salerno, zwischen partiellen und vollständigen Erklärungen, zwischen eigentlichen und uneigentlichen Erklärungen, gehe im Detail auf die These des Reduktionismus und auf unsere Erwartungen an die Wissenschaft ein und kann damit, anders als Salerno, mit zahlreichen Variationen des Gedankengangs, der auf Church zurückgeht, nicht nur gegen den Reduktionismus und gegen schwächere Versionen des Reduktionismus argumentieren; ich kann auch zeigen, was für eine Bedeutung dieser Gedankengang für unsere Erwartungen an die Wissenschaft hat.

[10]Es gibt schon allein deshalb Wahrheiten, die niemand vollständig erklärt (und die auch niemand jemals vollständig erklären wird), weil es Wahrheiten gibt, die niemanden interessieren (und die auch niemanden jemals interessieren werden). Ein kurzes Beispiel: Entweder die Quersumme der Anzahl der Haare auf meinem Kopf ist gerade eine Primzahl, dann wird niemand jemals vollständig erklären, warum die Quersumme der Anzahl der Haare auf meinem Kopf gerade eine Primzahl ist, oder die Quersumme der Anzahl der Haare auf meinem Kopf ist gerade keine Primzahl, dann wird niemand jemals vollständig erklären, warum die Quersumme der Anzahl der Haare auf meinem Kopf gerade keine Primzahl ist. Es gibt also eine Wahrheit, die niemand vollständig erklärt (und die auch niemand jemals vollständig erklären wird).

diese Wahrheit wahr ist und von niemandem vollständig erklärt wird, dass dann jemand vollständig erklären würde, warum diese Wahrheit wahr ist, und dass dann jemand vollständig erklären würde, warum diese Wahrheit von niemandem vollständig erklärt wird. Mein Argument setzt also das Konjunktionsprinzip voraus.

Mein Argument beruht zweitens auf der Annahme, dass man nur Wahrheiten vollständig erklären kann[11]:

Faktivitätsprinzip Wenn jemand vollständig erklärt, warum p, dann p.

Denn mein Argument setzt voraus, dass jemand nur vollständig erklären kann, warum eine Wahrheit von niemandem vollständig erklärt wird, wenn diese Wahrheit *tatsächlich* von niemandem vollständig erklärt wird. Mein Argument setzt also das Faktivitätsprinzip voraus.

Um mein Argument zurückzuweisen, müssen Reduktionisten also entweder das Konjunktionsprinzip, oder das Faktivitätsprinzip zurückweisen. Nun gibt es aber gute Gründe, das Konjunktionsprinzip für wahr zu halten. Ein Wissenschaftler, der behauptet, mit seiner Theorie *nicht* vollständig zu erklären, warum es Schwarze Löcher gibt, dafür aber *vollständig* zu erklären, warum es Schwarze Löcher und Dunkle Materie gibt, der hat offensichtlich nicht verstanden, was es bedeutet, *vollständig* zu erklären, warum eine Konjunktion wahr ist. Denn vollständig zu erklären, warum eine Konjunktion wahr ist, impliziert, vollständig zu erklären, warum alle Konjunkte dieser Konjunktion wahr sind. Wenn ein Wissenschaftler mit seiner Theorie nicht vollständig erklärt, warum es Schwarze Löcher gibt, dann erklärt er mit seiner Theorie auch *nicht* vollständig, warum es Schwarze Löcher und Dunkle Materie gibt. Er erklärt mit seiner Theorie, wenn überhaupt, *nur zum Teil,* warum es Schwarze Löcher und Dunkle Materie gibt. Es gibt also gute Gründe, das Konjunktionsprinzip für wahr zu halten.

Es gibt nicht nur gute Gründe, das Konjunktionsprinzip für wahr zu halten, es gibt auch gute Gründe, das Faktivitätsprinzip für wahr zu halten. Ein Wissenschaftler, der behauptet, dass sich Licht *nicht* im Äther ausbreitet, der aber gleichzeitig behauptet, mit seiner Theorie zu *erklären,* warum sich Licht im Äther ausbreitet, der hat offensichtlich nicht verstanden, was es bedeutet zu *erklären,* warum etwas wahr ist. Denn zu erklären, warum etwas wahr ist, impliziert, dass das, was man erklärt, auch *tatsächlich* wahr ist. Wenn sich Licht nicht im Äther ausbreitet, dann erklärt ein Wissenschaftler mit seiner Theorie auch nicht, warum sich Licht im Äther ausbreitet. Es gibt also nicht nur gute Gründe, das Konjunktionsprinzip für wahr zu halten, es gibt auch gute Gründe, das Faktivitätsprinzip für wahr zu halten.

Darüber hinaus gibt es gute Gründe, anzunehmen, dass das Konjunktionsprinzip nicht unbedingt nötig ist, um zu zeigen, dass es Wahrheiten gibt, die niemand vollständig erklären kann. Denn es genügt, erstens das Konjunktionsprinzip etwas abzuschwächen und zu behaupten, dass jemand eine Konjunktion nur vollständig erklärt, wenn er alle Konjunkte dieser Konjunktion *zumindest zum Teil* erklärt, zweitens das

---

[11]Zur Faktivität von Erklärungen vgl. Schnieder (2011, S. 451) und Fine (2012, S. 48–50).

Faktivitätsprinzip etwas umzuändern und zu behaupten, dass man nur Wahrheiten *zumindest zum Teil* erklären kann, und drittens anzunehmen, dass es (wenigstens) eine Wahrheit gibt, die von niemandem *zumindest zum Teil* erklärt wird (was plausibel ist, wenn man bedenkt, dass es überabzählbar viele Wahrheiten gibt). Es ist dann einfach, zu zeigen, dass es (wenigstens) eine Wahrheit gibt, die niemand vollständig erklären kann. Denn wenn es eine Wahrheit gibt, die von niemandem zumindest zum Teil erklärt wird, dann kann niemand vollständig erklären, warum diese Wahrheit wahr ist und von niemandem zumindest zum Teil erklärt wird. Denn wenn jemand vollständig erklären würde, warum diese Wahrheit wahr ist und von niemandem zumindest zum Teil erklärt wird, dann würde jemand (wenn das abgeschwächte Konjunktionsprinzip gültig ist) zumindest zum Teil erklären, warum diese Wahrheit wahr ist, und dann würde jemand zumindest zum Teil erklären, warum diese Wahrheit von niemandem zumindest zum Teil erklärt wird. Es wäre dann aber *sowohl wahr,* dass diese Wahrheit von jemandem zumindest zum Teil erklärt wird (denn dann würde jemand zumindest zum Teil erklären, warum diese Wahrheit wahr ist), *als auch nicht wahr,* dass diese Wahrheit von jemandem zumindest zum Teil erklärt wird (denn wenn das umgeänderte Faktivitätsprinzip gültig ist, dann kann man nur dann zumindest zum Teil erklären, warum diese Wahrheit von niemandem zumindest zum Teil erklärt wird, wenn diese Wahrheit *tatsächlich* von niemandem zumindest zum Teil erklärt wird). Es *kann* also niemand vollständig erklären, warum diese Wahrheit wahr ist und von niemandem zumindest zum Teil erklärt wird. Es gibt dann also Wahrheiten, die niemand vollständig erklären *kann.* Es ist also nicht unbedingt nötig, sich auf das Konjunktionsprinzip zu berufen, um zu zeigen, dass es Wahrheiten gibt, die niemand vollständig erklären kann.

Reduktionisten können meinem Argument deshalb, wenn überhaupt, nur entkommen, wenn sie die These des Reduktionismus abschwächen. Nach meiner Überzeugung kann man mithilfe modifizierter Versionen meines Arguments allerdings auch gegen schwächere Versionen des Reduktionismus argumentieren.

9.3.2 Nachtrag I: Es gibt in jedem Bereich unerklärliche Wahrheiten

Um zu zeigen, dass die These des Reduktionismus falsch ist, genügt es natürlich, zu zeigen, dass es *wenigstens eine* Wahrheit gibt, die nicht vollständig erklärt werden kann. Nun ist aber nicht auszuschließen, dass Reduktionisten in der Hoffnung, die These des Reduktionismus zu retten, zwischen einer stärkeren und einer schwächeren Version des Reduktionismus unterscheiden und sich auf die schwächere Version des Reduktionismus zurückziehen werden. Die These des *globalen Reduktionismus* wäre dieser Unterscheidung nach die These, dass *alle* Wahrheiten mithilfe von Wahrheiten über einen bestimmten Bereich vollständig erklärt werden können, die These des *lokalen Reduktionismus* wäre dagegen die These, dass lediglich alle Wahrheiten *über einen bestimmten Bereich* mithilfe von Wahrheiten über einen bestimmten Bereich vollständig erklärt werden können:

Globaler Reduktionismus *Alle* Wahrheiten können mithilfe von Wahrheiten über einen bestimmten Bereich vollständig erklärt werden.

Lokaler Reduktionismus Alle Wahrheiten *über einen bestimmten Bereich* können mithilfe von Wahrheiten über einen bestimmten Bereich vollständig erklärt werden.

Zur Veranschaulichung: Physikalistische Reduktionisten müssen nicht unbedingt behaupten, dass *alle* Wahrheiten mithilfe von Wahrheiten über physikalische Phänomene vollständig erklärt werden können. Sie können stattdessen behaupten, dass lediglich alle Wahrheiten *über mentale Phänomene* mithilfe von Wahrheiten über physikalische Phänomene vollständig erklärt werden können.[12] Sie wären dann nicht mehr Vertreter einer Version des globalen Reduktionismus, sondern nur noch Vertreter einer Version des lokalen Reduktionismus.

Nach meiner Auffassung ist allerdings nicht nur die These des globalen Reduktionismus, sondern auch die These des lokalen Reduktionismus zum Scheitern verurteilt. Denn der lokale Reduktionismus geht von einem bestimmten Bereich aus und setzt voraus, dass alle Wahrheiten über diesen Bereich vollständig erklärt werden können. Nach meiner Auffassung gibt es aber in *jedem* Bereich Wahrheiten, die nicht vollständig erklärt werden können. Um nachvollziehen zu können, warum es in jedem Bereich Wahrheiten gibt, die nicht vollständig erklärt werden können, gehen wir von einem beliebigen Bereich aus. Die Konjunktion einer Wahrheit über diesen Bereich mit einer beliebigen Wahrheit ist selbst eine Wahrheit über diesen Bereich.[13] Es gibt aber, wie wir bereits gesehen haben, überabzählbar viele Wahrheiten. Es gibt also überabzählbar viele Konjunktionen einer Wahrheit über diesen Bereich mit einer beliebigen Wahrheit. Es gibt also überabzählbar viele Wahrheiten über diesen Bereich. Was für einen beliebigen Bereich gilt, gilt für jeden Bereich: Es gibt überabzählbar viele Wahrheiten über jeden Bereich.

Nicht nur (aber auch) aus diesem Grund ist es plausibel anzunehmen, dass es in jedem Bereich (wenigstens) eine Wahrheit gibt, die niemand vollständig erklärt. Es folgt dann aber, dass es in jedem Bereich (wenigstens) eine Wahrheit gibt, die niemand vollständig erklären *kann*. Denn gehen wir einmal von einem beliebigen Bereich aus und nehmen an, jemand würde vollständig erklären, warum die Wahrheit über diesen Bereich, die niemand vollständig erklärt, wahr ist und von niemandem vollständig erklärt wird. Wenn jemand vollständig erklären würde, warum diese Wahrheit wahr ist und von niemandem vollständig erklärt wird, dann wäre es, wie

---

[12]Der physikalistische Reduktionismus würde dann, im Unterschied zum physikalistischen Eliminativismus, zugeben, dass es mentale Phänomene und damit Wahrheiten über mentale Phänomene gibt; er würde aber behaupten, dass man alle Wahrheiten über mentale Phänomene auf Wahrheiten über physikalische Phänomene reduzieren kann.

[13]Die Wahrheit, dass Quarks eine Spinquantenzahl haben und dass Napoleon die Schlacht von Waterloo im Jahr 1815 verloren hat, ist eine Wahrheit über physikalische Phänomene (weil die Wahrheit, dass Quarks eine Spinquantenzahl haben, eine Wahrheit über physikalische Phänomene ist). Die Wahrheit, dass Leibniz und Newton die Infinitesimalrechnung entwickelt haben und dass π eine irrationale Zahl ist, ist eine Wahrheit über mentale Phänomene (weil die Wahrheit, dass Leibniz und Newton die Infinitesimalrechnung entwickelt haben, eine Wahrheit über mentale Phänomene ist).

wir bereits gesehen haben, *sowohl wahr,* dass diese Wahrheit von jemandem vollständig erklärt wird (denn dann würde jemand vollständig erklären, warum diese Wahrheit wahr ist), *als auch nicht wahr,* dass diese Wahrheit von jemandem vollständig erklärt wird (denn jemand kann nur dann vollständig erklären, warum diese Wahrheit von niemandem vollständig erklärt wird, wenn diese Wahrheit *tatsächlich* von niemandem vollständig erklärt wird). Es *kann* also niemand vollständig erklären, warum diese Wahrheit wahr ist und von niemandem vollständig erklärt wird. Nun ist diese Wahrheit aber eine Wahrheit über den Bereich, von dem wir ausgegangen sind, und die Konjunktion einer Wahrheit über diesen Bereich mit einer beliebigen Wahrheit ist selbst eine Wahrheit über diesen Bereich. Es ist also eine Wahrheit über diesen Bereich, dass diese Wahrheit wahr ist und von niemandem vollständig erklärt wird. Nun kann aber, wie wir gesehen haben, niemand vollständig erklären, warum diese Wahrheit wahr ist und von niemandem vollständig erklärt wird. Es gibt also eine Wahrheit über diesen Bereich, die niemand vollständig erklären kann. Und was für einen beliebigen Bereich gilt, gilt für jeden Bereich: Es gibt in jedem Bereich (wenigstens) eine Wahrheit, die niemand vollständig erklären kann.

Eine formallogische Ausarbeitung dieses Arguments befindet sich in Abschn. 9.6. Es würde also nichts bringen, zwischen einer globalen und einer lokalen Version des Reduktionismus zu unterscheiden und sich auf die lokale Version des Reduktionismus zurückzuziehen. Denn die lokale Version des Reduktionismus ist ebenfalls zum Scheitern verurteilt.

9.3.3 Nachtrag II: Es gibt völlig unerklärliche Wahrheiten

Um zu zeigen, dass die These des Reduktionismus falsch ist, genügt es natürlich, zu zeigen, dass es Wahrheiten gibt, die nicht *vollständig* erklärt werden können. Es ist aber nicht auszuschließen, dass Reduktionisten in der Hoffnung, die These des Reduktionismus zu retten, *in einem anderen Sinn* zwischen einer stärkeren und einer schwächeren Version des Reduktionismus unterscheiden und sich auf die schwächere Version des Reduktionismus zurückziehen werden. Die These des *starken Reduktionismus* wäre dieser Unterscheidung nach die These, dass alle Wahrheiten mithilfe von Wahrheiten über einen bestimmten Bereich *vollständig* erklärt werden können; die These des *schwachen Reduktionismus* wäre dagegen die These, dass alle Wahrheiten mithilfe von Wahrheiten über einen bestimmten Bereich *zumindest zum Teil* erklärt werden können:

Starker Reduktionismus Alle Wahrheiten können mithilfe von Wahrheiten über einen bestimmten Bereich *vollständig* erklärt werden.

Schwacher Reduktionismus Alle Wahrheiten können mithilfe von Wahrheiten über einen bestimmten Bereich *zumindest zum Teil* erklärt werden.

Nach meiner Überzeugung würde der schwache Reduktionismus aber nicht nur den Ansprüchen nicht gerecht werden, die man für gewöhnlich mit dem Reduktionismus verbindet; der schwache Reduktionismus wäre genauso wie der starke Reduktionismus zum Scheitern verurteilt. Denn der schwache Reduktionismus würde voraussetzen, dass man alle Wahrheiten *zumindest zum Teil* erklären kann. Nach meiner Auffassung kann man den Gedankengang, den Church entwickelt hat, aber nicht nur aufgreifen und (leicht) modifizieren, um zu zeigen, dass es Wahrheiten gibt, die *nicht vollständig* erklärt werden können; man kann ihn auch aufgreifen und (etwas stärker) modifizieren, um zu zeigen, dass es Wahrheiten gibt, die *nicht einmal zum Teil* erklärt werden können.[14]

Williamson (2000, S. 283 f.) hat in Anlehnung an den Gedankengang, der auf Church zurückgeht, einen etwas modifizierten Gedankengang entwickelt, um mit etwas anderen Annahmen zu zeigen, dass es Wahrheiten gibt, die niemand wissen kann. Um Williamsons Gedankengang darzustellen, ist es hilfreich, drei Definitionen einzuführen: Eine Wahrheit ist genau dann *bekannt,* wenn jemand weiß, dass diese Wahrheit wahr ist, eine Wahrheit ist genau dann *unbekannt,* wenn niemand weiß, dass diese Wahrheit wahr ist, und eine Wahrheit ist genau dann *völlig unbekannt,* wenn diese Wahrheit nicht Konjunkt einer bekannten Konjunktion ist. Williamson geht davon aus, dass es (wenigstens) eine völlig unbekannte Wahrheit gibt. Wenn es aber (wenigstens) eine völlig unbekannte Wahrheit gibt, dann kann niemand wissen, dass diese Wahrheit wahr und völlig unbekannt ist. Denn wenn jemand wüsste, dass diese Wahrheit wahr und völlig unbekannt ist, dann wäre es *sowohl wahr,* dass diese Wahrheit völlig unbekannt ist (denn niemand kann wissen, dass diese Wahrheit wahr und völlig unbekannt ist, wenn diese Wahrheit nicht *tatsächlich* wahr und völlig unbekannt ist), *als auch nicht wahr,* dass diese Wahrheit völlig unbekannt ist (denn wenn jemand weiß, dass diese Wahrheit wahr und völlig unbekannt ist, dann ist diese Wahrheit Konjunkt einer bekannten Konjunktion und folglich nicht völlig unbekannt). Es *kann* also niemand wissen, dass diese Wahrheit wahr und völlig unbekannt ist. Es gibt also Wahrheiten, die niemand wissen kann, wenn es Wahrheiten gibt, die völlig unbekannt sind. Nun gibt es nach Williamson aber Wahrheiten, die völlig unbekannt sind. Es gibt also Wahrheiten, die niemand wissen kann.

Nach meiner Auffassung kann man in Anlehnung an Williamsons Gedankengang dafür argumentieren, dass es Wahrheiten gibt, die *nicht einmal zum Teil* erklärt werden können. Denn es ist klar, dass nicht alle Wahrheiten erwähnt werden (schon allein deshalb, weil es überabzählbar viele Wahrheiten gibt). Wenn es Wahrheiten gibt, die

---

[14]Der Grund, warum es einer Modifikation bedarf, um zu zeigen, dass es Wahrheiten gibt, die nicht einmal zum Teil erklärt werden können, ist, dass das Konjunktionsprinzip, das für *vollständige* Erklärungen gültig ist, für *partielle* Erklärungen ungültig ist. Es ist zum Beispiel möglich, dass *jemand zum Teil erklärt,* warum sich jede positive ganze Zahl als Produkt von Primzahlen darstellen lässt und warum das Licht, das von einer Galaxie ausgeht, ein bestimmtes Verhältnis von Wellenlängenänderung und ursprünglicher Wellenlänge aufweist (weil jemand erklärt, warum sich jede positive ganze Zahl als Produkt von Primzahlen darstellen lässt), dass aber *niemand zum Teil erklärt,* warum das Licht, das von einer Galaxie ausgeht, ein bestimmtes Verhältnis von Wellenlängenänderung und ursprünglicher Wellenlänge aufweist. Das Konjunktionsprinzip, das für vollständige Erklärungen gültig ist, ist also für partielle Erklärungen ungültig. Zur partiellen Erklärung von Konjunktionen durch ihre Konjunkte vgl. Schnieder (2011, S. 454).

nicht erwähnt werden, dann gibt es aber auch Wahrheiten, die nicht *im Rahmen einer Erklärung* erwähnt werden.

Es ist nun nicht schwierig, zu zeigen, dass es Wahrheiten gibt, die nicht einmal zum Teil erklärt werden können. Denn wenn es eine Wahrheit gibt, die nicht im Rahmen einer Erklärung erwähnt wird, dann kann niemand zumindest zum Teil erklären, warum diese Wahrheit wahr ist und nicht im Rahmen einer Erklärung erwähnt wird. Denn wenn jemand zumindest zum Teil erklären würde, warum diese Wahrheit wahr ist und nicht im Rahmen einer Erklärung erwähnt wird, dann wäre es *sowohl wahr,* dass diese Wahrheit im Rahmen einer Erklärung erwähnt wird (denn dann würde jemand erklären, warum diese Wahrheit wahr ist und nicht im Rahmen einer Erklärung erwähnt wird), *als auch nicht wahr,* dass diese Wahrheit im Rahmen einer Erklärung erwähnt wird (denn niemand kann zumindest zum Teil erklären, warum diese Wahrheit wahr ist und nicht im Rahmen einer Erklärung erwähnt wird, wenn diese Wahrheit nicht *tatsächlich* nicht im Rahmen einer Erklärung erwähnt wird). Es *kann* also niemand zumindest zum Teil erklären, warum diese Wahrheit wahr ist und nicht im Rahmen einer Erklärung erwähnt wird. Es gibt also Wahrheiten, die niemand zumindest zum Teil erklären kann, wenn es Wahrheiten gibt, die nicht im Rahmen einer Erklärung erwähnt werden. Nun gibt es aber Wahrheiten, die nicht im Rahmen einer Erklärung erwähnt werden. Es gibt also Wahrheiten, die *nicht einmal zum Teil* erklärt werden können.

Eine formallogische Ausarbeitung meines Arguments befindet sich in Abschn. 9.6. Es würde also nichts bringen, zwischen einer stärkeren und einer schwächeren Version des Reduktionismus zu unterscheiden und sich auf die schwächere Version des Reduktionismus zurückzuziehen. Denn die schwächere Version des Reduktionismus ist ebenfalls zum Scheitern verurteilt.[15]

9.4   Konsequenzen für unsere Erwartungen an die Wissenschaft

Wenn eine Wahrheit genau dann *unerklärlich* ist, wenn niemand diese Wahrheit vollständig erklären kann, und eine Wahrheit genau dann *unerklärt* ist, wenn niemand diese Wahrheit vollständig erklärt, dann können wir festhalten: Es ist unerklärlich, warum unerklärte Wahrheiten wahr und unerklärt sind. Es gibt also unerklärliche Wahrheiten. Wenn eine Wahrheit außerdem genau dann *völlig unerklärlich* ist, wenn man diese Wahrheit nicht einmal zum Teil erklären kann, und eine Wahrheit genau dann *unerwähnt* ist, wenn diese Wahrheit nicht im Rahmen einer Erklärung erwähnt

---

[15]Es ist nicht schwierig zu zeigen, dass eine noch schwächere Version des Reduktionismus (eine Kombination von lokalem und schwachem Reduktionismus, wonach alle Wahrheiten *über einen bestimmten Bereich* mithilfe von Wahrheiten über einen bestimmten Bereich *zumindest zum Teil* erklärt werden können) ebenfalls zum Scheitern verurteilt ist. Hierfür genügt es, das Argument gegen den lokalen Reduktionismus und das Argument gegen den schwachen Reduktionismus auf naheliegende Weise zu kombinieren.

wird, dann können wir hinzufügen: Es ist völlig unerklärlich, warum unerwähnte Wahrheiten wahr und unerwähnt sind. Es gibt also völlig unerklärliche Wahrheiten. Es gibt also unerklärliche Wahrheiten. Wenn der Reduktionismus wahr ist, dann gibt es aber keine unerklärlichen Wahrheiten. Denn wenn der Reduktionismus wahr ist, dann kann man alle Wahrheiten vollständig erklären. Der Reduktionismus ist also falsch.

Der Traum von der vollständigen Erklärung der Welt durch die Wissenschaft ist also aus prinzipiellen Gründen zum Scheitern verurteilt. Das Problem ist nicht nur, dass es zu viele Wahrheiten gibt (und wir nie an ein Ende kommen würden, wenn wir versuchen würden, alle Wahrheiten einzeln aufzuzählen und zu erklären). Das Problem ist vielmehr, dass nicht alle Wahrheiten vollständig erklärt werden können. Egal wie viele Wahrheiten wir aufzählen würden: Mithilfe dieser Wahrheiten können wir aus prinzipiellen Gründen niemals *alle* Wahrheiten vollständig erklären.

Der Traum von der vollständigen Erklärung der Welt durch die Wissenschaft ist also zum Scheitern verurteilt. Dennoch könnte man versuchen, diesen Traum zumindest in abgeschwächter Form aufrechtzuerhalten: Man müsste zwar zugeben, dass es Wahrheiten gibt, die nicht vollständig erklärt werden können,[16] und damit zugeben, dass man die Welt niemals in dem Sinn vollständig erklären können wird, dass man endlich viele Wahrheiten aufzählt, mit denen man dann zumindest im Prinzip *alle* Wahrheiten vollständig erklären kann. Dennoch könnte man hoffen, die Welt irgendwann einmal zumindest in dem Sinn vollständig erklären zu können, dass man endlich viele Wahrheiten aufzählt, mit denen man dann zumindest im Prinzip *alle anderen* Wahrheiten vollständig erklären kann. Hierfür, so der Vorschlag, müsste man lediglich alle unerklärlichen Wahrheiten ausfindig machen, aufzählen und dann lediglich *alle anderen* Wahrheiten erklären können. Die Behauptung wäre dann nicht mehr, mithilfe der aufgezählten Wahrheiten *alle* Wahrheiten erklären zu können (denn hierfür müsste man auch die unerklärlichen Wahrheiten erklären). Die Behauptung wäre lediglich, mithilfe der aufgezählten Wahrheiten *alle anderen* Wahrheiten erklären zu können (denn dann müsste man nicht die aufgezählten Wahrheiten und damit auch nicht die unerklärlichen Wahrheiten unter den aufgezählten Wahrheiten erklären). Man scheint dann den Traum von der vollständigen Erklärung der Welt durch die Wissenschaft zumindest in einem gewissen Sinn weiter aufrechterhalten zu können.

Dieser Vorschlag ist nach meiner Auffassung allerdings ebenfalls zum Scheitern verurteilt. Denn um diesen Traum zu verwirklichen, müsste man alle unerklärlichen Wahrheiten ausfindig machen und aufzählen. Es ist aber plausibel anzunehmen, dass wir niemals alle unerklärlichen Wahrheiten ausfindig machen werden, und es gibt gute Gründe anzunehmen, dass es unmöglich ist, alle unerklärlichen Wahrheiten aufzuzählen. Es gibt in meinen Augen deshalb zwei Gründe, anzunehmen, dass dieser Traum nicht verwirklicht werden kann:

---

[16]Manch einer wird behaupten, dass mein Ergebnis (zumindest mein Ergebnis, dass es Wahrheiten gibt, die man nicht vollständig erklären kann) ohnehin nicht überraschend ist – zumindest dann nicht, wenn man bedenkt, dass es Gründe gibt, anzunehmen, dass es physikalische Wahrheiten gibt, die man nicht vollständig erklären kann (zum Beispiel physikalische Wahrheiten über den Zerfall eines Radiumatoms).

1. Es ist plausibel anzunehmen, dass wir niemals alle unerklärlichen Wahrheiten ausfindig machen werden. Denn wir werden erst dann alle unerklärlichen Wahrheiten ausfindig gemacht haben, wenn wir allwissend geworden sind (und es ist plausibel anzunehmen, dass wir niemals allwissend sein werden). Um zu zeigen, dass wir erst dann alle unerklärlichen Wahrheiten ausfindig gemacht haben werden, wenn wir allwissend geworden sind, nehmen wir an, wir hätten alle unerklärlichen Wahrheiten ausfindig gemacht. Wenn wir alle unerklärlichen Wahrheiten ausfindig gemacht hätten, dann würden wir alle unerklärlichen Wahrheiten kennen (wir wüssten dann von allen unerklärlichen Wahrheiten, dass sie wahr und unerklärlich sind). Es ist aber nicht schwer, zu zeigen, dass wir dann nicht nur *alle unerklärlichen* Wahrheiten, sondern *alle* Wahrheiten kennen würden (dass wir dann von *allen* Wahrheiten wissen würden, dass sie wahr sind). Um nachvollziehen zu können, warum wir dann nicht nur alle unerklärlichen Wahrheiten, sondern alle Wahrheiten kennen würden, ist es wichtig, sich klarzumachen, dass die Konjunktion einer unerklärlichen Wahrheit mit einer beliebigen Wahrheit selbst eine unerklärliche Wahrheit ist. Denn wenn die Konjunktion einer unerklärlichen Wahrheit mit einer beliebigen Wahrheit selbst *keine* unerklärliche Wahrheit wäre, dann gäbe es eine Konjunktion, die man vollständig erklären kann, obwohl man *nicht* alle Konjunkte dieser Konjunktion vollständig erklären kann. Wie wir bereits gesehen haben, kann man eine Konjunktion aber nur vollständig erklären, wenn man alle Konjunkte dieser Konjunktion vollständig erklären kann. Die Konjunktion einer unerklärlichen Wahrheit mit einer beliebigen Wahrheit ist also selbst eine unerklärliche Wahrheit. Nun gehen wir aber von einer beliebigen Wahrheit aus und nehmen an, wir würden alle unerklärlichen Wahrheiten kennen. Die Konjunktion dieser beliebigen Wahrheit mit einer unerklärlichen Wahrheit wäre, wie wir gesehen haben, selbst unerklärlich. Wir *wüssten* dann, da wir alle unerklärlichen Wahrheiten kennen, dass die Konjunktion dieser beliebigen Wahrheit mit einer unerklärlichen Wahrheit wahr ist. Wir *wüssten* dann aber auch, dass diese Wahrheit wahr ist (denn wer weiß, dass eine Konjunktion wahr ist, weiß auch, dass alle Konjunkte dieser Konjunktion wahr sind). Wir würden also, wenn wir alle unerklärlichen Wahrheiten kennen würden, jede beliebige Wahrheit kennen. Wir würden also, wenn wir alle unerklärlichen Wahrheiten kennen würden, *alle* Wahrheiten kennen. Wir wären also, wenn wir alle unerklärlichen Wahrheiten ausfindig gemacht hätten, allwissend. Nun ist es aber plausibel anzunehmen, dass wir niemals allwissend sein werden. Es ist also auch plausibel anzunehmen, dass wir niemals alle unerklärlichen Wahrheiten ausfindig machen werden. Eine formallogische Ausarbeitung des wichtigsten Teils meines Arguments befindet sich in Abschn. 9.6.

2. Es gibt gute Gründe anzunehmen, dass es unmöglich ist, alle unerklärlichen Wahrheiten aufzuzählen. Denn wir können alle unerklärlichen Wahrheiten nur aufzählen, wenn es *endlich viele* unerklärlichen Wahrheiten gibt. Es gibt aber überabzählbar viele unerklärlichen Wahrheiten. Denn, wie wir bereits gesehen haben, ist die Konjunktion einer unerklärlichen Wahrheit mit einer beliebigen Wahrheit selbst eine unerklärliche Wahrheit. Nun gibt es aber überabzählbar viele Wahrheiten. Es gibt also überabzählbar viele Konjunktionen einer unerklärlichen Wahrheit mit

einer beliebigen Wahrheit. Es gibt also überabzählbar viele unerklärliche Wahrheiten. Wir können also nicht alle unerklärlichen Wahrheiten aufzählen.

Der Traum, die Welt wenigstens in dem Sinn vollständig zu erklären, dass man
endlich viele Wahrheiten aufzählt, mit denen man dann zumindest im Prinzip *alle*
anderen Wahrheiten vollständig erklären kann, ist also ebenfalls zum Scheitern verurteilt. Denn hierfür müsste man alle unerklärlichen Wahrheiten ausfindig machen
und aufzählen. Es ist aber plausibel anzunehmen, dass wir niemals alle unerklärlichen Wahrheiten ausfindig machen werden, und es gibt gute Gründe anzunehmen,
dass es unmöglich ist, alle unerklärlichen Wahrheiten aufzuzählen.

9.5   Fazit

Die These des Reduktionismus ist die These, dass alle Wahrheiten mithilfe von Wahrheiten über einen bestimmten Bereich vollständig erklärt werden können. Die These
des Reduktionismus setzt voraus, dass alle Wahrheiten vollständig erklärt werden
können. Nun gibt es aber (wenigstens) eine unerklärte Wahrheit (eine Wahrheit, die
niemand vollständig erklärt). Es gibt also (wenigstens) eine Frage, die Reduktionisten beantworten müssten (wenn der Reduktionismus wahr wäre), aber nicht können
(da es widersprüchlich ist, anzunehmen, dass sie diese Frage beantworten) – die
Frage, warum diese Wahrheit wahr und unerklärt ist. Denn wenn Reduktionisten
die Frage, warum diese Wahrheit wahr und unerklärt ist, beantworten würden, dann
würden sie erklären, warum diese Wahrheit wahr und unerklärt ist. Dann wäre es aber
sowohl wahr als auch nicht wahr, dass diese Wahrheit unerklärt ist (was unmöglich
ist). Es gibt also (wenigstens) eine Frage, die Reduktionisten beantworten müssten,
aber nicht können – die Frage, warum diese Wahrheit wahr und unerklärt ist. Der
Reduktionismus ist also aus prinzipiellen Gründen falsch.

Wenn der Reduktionismus falsch ist, dann wird die Wissenschaft die Welt allerdings niemals vollständig erklären können. Denn wenn es wenigstens eine Wahrheit
gibt, die wir nicht vollständig erklären können, dann ist der Versuch, *alle* Wahrheiten über die Welt vollständig zu erklären, völlig aussichtslos. Und wenn es *eine*
Wahrheit gibt, die wir nicht vollständig erklären können, dann gibt es *überabzählbar*
viele Wahrheiten, die wir nicht vollständig erklären können. Aus diesem Grund ist
der Versuch, alle Wahrheiten über die Welt, die wir nicht vollständig erklären können, aufzuzählen und nur *alle anderen* Wahrheiten vollständig zu erklären, ebenfalls
völlig aussichtslos. Die Wissenschaft wird die Welt niemals vollständig erklären
können.

9.6 Anhang: Formallogische Ausarbeitung

9.6.1 Es gibt unerklärliche Wahrheiten

Um in einer formallogischen Sprache zu zeigen, dass es unerklärliche Wahrheiten gibt, bedarf es zunächst der Einführung einiger Abkürzungen: Wenn A eine Formel der formallogischen Sprache ist, $\ulcorner \lozenge A \urcorner$ eine Abkürzung für $\ulcorner$ es ist möglich, dass $A \urcorner$, $\ulcorner \square A \urcorner$ eine Abkürzung für $\ulcorner$ es ist notwendig, dass $A \urcorner$ und $\ulcorner \mathcal{E} A \urcorner$ eine Abkürzung für $\ulcorner$ es gibt jemanden, der vollständig erklärt, warum $A \urcorner$, dann genügt es, wie in der Modallogik üblich, anzunehmen, dass die folgenden logischen Prinzipien gültig sind (Vgl. z. B. Kripke 1959, S. 1):

$$(\text{DEF-}\lozenge) \quad \vdash \lozenge A \leftrightarrow {\sim}\square{\sim}A$$
$$(\text{RN}) \quad \text{Wenn} \vdash A, \text{dann} \vdash \square A$$

Wenn p eine propositionale Variable und $A[B/p]$ die Formel ist, die dadurch entsteht, dass man alle freien Vorkommnisse von p in A mit der Formel B ersetzt (ohne dass dadurch eine freie Variable in B gebunden wird), dann kann man außerdem, wie in einer Logik mit propositionalen Quantoren üblich, annehmen, dass gilt (Vgl. z. B. Williamson 2007, S. 296):

$$(\text{ELIM-}\forall) \quad \vdash \forall p A \to A[B/p]$$
(INTRO-∀)   Wenn p eine propositionale Variable ist, die in A nicht frei ist,
            dann gilt: Wenn $\vdash A \to B$, dann $\vdash A \to \forall p B$
$$(\text{DEF-}\exists) \quad \vdash \exists p A \leftrightarrow {\sim}\forall p {\sim}A$$

Um zu zeigen, dass es unerklärliche Wahrheiten gibt, bedarf es außerdem zwei grundlegender erkenntnistheoretischer Prinzipien.

1. Wenn jemand vollständig erklärt, warum $A \& B$, dann erklärt jemand vollständig, warum A, und dann erklärt jemand vollständig, warum B:

$$(\text{KON-}\mathcal{E}) \vdash \mathcal{E}(A \& B) \to (\mathcal{E} A \& \mathcal{E} B)$$

2. Wenn jemand vollständig erklärt, warum A, dann A:

$$(\text{FAKT-}\mathcal{E}) \quad \vdash \mathcal{E} A \to A$$

Um zu zeigen, dass es unerklärliche Wahrheiten gibt, genügt es dann, davon auszugehen, dass es (wenigstens) eine Wahrheit gibt, die nicht vollständig von jemandem erklärt wird:

$$\exists p(p \& {\sim}\mathcal{E} p) \qquad \text{(Prämisse)} \qquad (9.1)$$

Denn wenn wir annehmen, dass alle Wahrheiten von jemandem vollständig erklärt werden können, dann können wir folgern, dass dann auch von jemandem vollständig erklärt werden kann, warum diese Wahrheit wahr ist und von niemandem vollständig erklärt wird:

$$\forall p(p \to \Diamond\mathcal{E}p) \qquad\qquad \text{(Annahme)} \qquad (9.2)$$

$$p \,\&\, {\sim}\mathcal{E} \qquad\qquad \text{(Annahme)} \qquad (9.3)$$

$$p \,\&\, {\sim}\mathcal{E}p \to \Diamond\mathcal{E}(p \,\&\, {\sim}\mathcal{E}p) \qquad\qquad \text{(Gl. 9.2, ELIM-}\forall\text{)} \qquad (9.4)$$

$$\Diamond\mathcal{E}(p \,\&\, {\sim}\mathcal{E}p) \qquad\qquad \text{(Gl. 9.3, Gl. 9.4)} \qquad (9.5)$$

Es ist also möglich, dass jemand vollständig erklärt, warum diese Wahrheit wahr ist und von niemandem vollständig erklärt wird. Nun nehmen wir aber einmal an, jemand würde vollständig erklären, warum diese Wahrheit wahr ist und von niemandem vollständig erklärt wird:

$$\mathcal{E}(p \,\&\, {\sim}\mathcal{E}p) \qquad\qquad \text{(Annahme)} \qquad (9.6)$$

Jemand würde dann vollständig erklären, warum diese Wahrheit wahr ist, und jemand würde dann vollständig erklären, warum diese Wahrheit von niemandem vollständig erklärt wird:

$$\mathcal{E}p \,\&\, \mathcal{E}{\sim}\mathcal{E}p \qquad\qquad \text{(Gl. 9.6, KON-}\mathcal{E}\text{)} \qquad (9.7)$$

Es wäre dann aber sowohl wahr, dass jemand vollständig erklärt, warum diese Wahrheit wahr ist, als auch nicht wahr, dass jemand vollständig erklärt, warum diese Wahrheit wahr ist:

$$\mathcal{E}p \,\&\, {\sim}\mathcal{E}p \qquad\qquad \text{(Gl. 9.7, FAKT-}\mathcal{E}\text{)} \qquad (9.8)$$

Es ist also widersprüchlich anzunehmen, jemand würde vollständig erklären, warum diese Wahrheit wahr ist und von niemandem vollständig erklärt wird. Es folgt, dass notwendig ist, dass niemand vollständig erklärt, warum diese Wahrheit wahr ist und von niemandem vollständig erklärt wird:

$$\Box{\sim}\mathcal{E}(p \,\&\, {\sim}\mathcal{E}p) \qquad\qquad \text{(Gl. 9.6–9.8, RN)} \qquad (9.9)$$

Es ist dann aber unmöglich, dass jemand vollständig erklärt, warum diese Wahrheit wahr ist und von niemandem vollständig erklärt wird:

$$\sim\!\Diamond\mathcal{E}(p \,\&\, {\sim}\mathcal{E}p) \qquad\qquad \text{(Gl. 9.9, DEF-}\Diamond\text{)} \qquad (9.10)$$

Es ist also widersprüchlich anzunehmen, dass diese Wahrheit wahr ist und von niemandem vollständig erklärt wird. Denn wenn diese Wahrheit wahr ist und von niemandem vollständig erklärt wird, dann ist es möglich und unmöglich, dass jemand vollständig erklärt, warum diese Wahrheit wahr ist und von niemandem vollständig erklärt wird:

$$\Diamond \mathcal{E}(p \,\&\, {\sim}\mathcal{E}p) \,\&\, {\sim}\Diamond \mathcal{E}(p \,\&\, {\sim}\mathcal{E}p) \qquad \text{(Gl. 9.5 und 9.10)} \qquad (9.11)$$

Es folgt, dass es falsch ist, dass diese Wahrheit wahr ist und von niemandem vollständig erklärt wird, wenn alle Wahrheiten von jemandem vollständig erklärt werden können:

$$\forall p(p \rightarrow \Diamond \mathcal{E}p) \rightarrow {\sim}(p \,\&\, {\sim}\mathcal{E}p) \qquad \text{(Gl. 9.2–9.11)} \qquad (9.12)$$

Es folgt, dass nicht alle Wahrheiten von jemandem vollständig erklärt werden können, wenn es (wenigstens) eine Wahrheit gibt, die nicht von jemandem vollständig erklärt wird:

$$\forall p(p \rightarrow \Diamond \mathcal{E}p) \rightarrow \forall p{\sim}(p \,\&\, {\sim}\mathcal{E}p) \qquad \text{(Gl. 9.2–9.12, INTRO-}\forall) \qquad (9.13)$$

$${\sim}\forall p{\sim}(p \,\&\, {\sim}\mathcal{E}p) \rightarrow {\sim}\forall p(p \rightarrow \Diamond \mathcal{E}p) \qquad \text{(Gl. 9.13)} \qquad (9.14)$$

$$\exists p(p \,\&\, {\sim}\mathcal{E}p) \rightarrow {\sim}\forall p(p \rightarrow \Diamond \mathcal{E}p) \qquad \text{(Gl. 9.14, DEF-}\exists) \qquad (9.15)$$

Nun gibt es aber (wenigstens) eine Wahrheit, die nicht vollständig von jemandem erklärt wird. Es folgt, dass nicht alle Wahrheiten von jemandem vollständig erklärt werden können:

$${\sim}\forall p(p \rightarrow \Diamond \mathcal{E}p) \qquad \text{(Gl. 9.1 und 9.15)} \qquad (9.16)$$

Es gibt also wenigstens eine Wahrheit, die nicht von jemandem vollständig erklärt werden kann:

$$\exists p(p \,\&\, {\sim}\Diamond \mathcal{E}p) \qquad \text{(Gl. 9.16, DEF-}\exists) \qquad (9.17)$$

Es gibt also wenigstens eine unerklärliche Wahrheit. QED.

9.6.2 Es gibt in jedem Bereich unerklärliche Wahrheiten

Um in einer formallogischen Sprache zu zeigen, dass es in jedem Bereich unerklärliche Wahrheiten gibt, genügt es, eine zusätzliche Abkürzung und eine zusätzliche Annahme einzuführen: Wenn $\ulcorner \beta A \urcorner$ eine Abkürzung für $\ulcorner$es ist eine Wahrheit über den Bereich β, dass $A \urcorner$ ist, dann genügt es vorauszusetzen, dass die Konjunktion einer Wahrheit über den Bereich β mit einer beliebigen Wahrheit selbst eine Wahrheit über den Bereich β ist:

$$\text{(KON-}\beta) \qquad \vdash \beta A \rightarrow (B \rightarrow \beta(A \,\&\, B))$$

Um zu zeigen, dass es in jedem Bereich unerklärliche Wahrheiten gibt, genügt es dann, von einem beliebigen Bereich β auszugehen und anzunehmen, dass es (wenigstens) eine Wahrheit über den Bereich β gibt, die nicht vollständig von jemandem erklärt wird:

$$\exists p(\beta p \,\&\, \sim\!\mathcal{E} p) \qquad\qquad \text{Prämisse} \qquad\qquad (9.18)$$

Denn wenn wir annehmen, dass alle Wahrheiten über den Bereich β von jemandem vollständig erklärt werden können, dann können wir folgern, dass dann auch von jemandem vollständig erklärt werden kann, warum diese Wahrheit wahr ist und von niemandem vollständig erklärt wird:

$$\forall p(\beta p \rightarrow \Diamond\mathcal{E} p) \qquad\qquad \text{(Annahme)} \qquad\qquad (9.19)$$

$$\beta p \,\&\, \sim\!\mathcal{E} p \qquad\qquad \text{(Annahme)} \qquad\qquad (9.20)$$

$$\beta(p \,\&\, \sim\!\mathcal{E} p)) \qquad\qquad \text{(Gl. 9.20, KON-}\beta\text{)} \qquad\qquad (9.21)$$

$$\beta(p \,\&\, \sim\!\mathcal{E} p) \rightarrow \Diamond\mathcal{E}(p \,\&\, \sim\!\mathcal{E} p) \qquad \text{(Gl. 9.19, ELIM-}\forall\text{)} \qquad (9.22)$$

$$\Diamond\mathcal{E}(p \,\&\, \sim\!\mathcal{E} p) \qquad\qquad \text{(Gl. 9.21 und 9.22)} \qquad\qquad (9.23)$$

Es ist also möglich, dass jemand vollständig erklärt, warum diese Wahrheit wahr ist und von niemandem vollständig erklärt wird. Nun nehmen wir aber einmal an, jemand würde vollständig erklären, warum diese Wahrheit wahr ist und von niemandem vollständig erklärt wird:

$$\mathcal{E}(p \,\&\, \sim\!\mathcal{E} p) \qquad\qquad \text{(Annahme)} \qquad\qquad (9.24)$$

Jemand würde dann vollständig erklären, warum diese Wahrheit wahr ist, und jemand würde dann vollständig erklären, warum diese Wahrheit von niemandem vollständig erklärt wird:

$$\mathcal{E} p \,\&\, \mathcal{E}\!\sim\!\mathcal{E} p \qquad\qquad \text{(Gl. 9.24, KON-}\mathcal{E}\text{)} \qquad\qquad (9.25)$$

Es wäre dann aber sowohl wahr, dass jemand vollständig erklärt, warum diese Wahrheit wahr ist, als auch nicht wahr, dass jemand vollständig erklärt, warum diese Wahrheit wahr ist:

$$\mathcal{E} p \,\&\, \sim\!\mathcal{E} p \qquad\qquad \text{(Gl. 9.25, FAKT-}\mathcal{E}\text{)} \qquad\qquad (9.26)$$

Es ist also widersprüchlich anzunehmen, jemand würde vollständig erklären, warum diese Wahrheit wahr ist und von niemandem vollständig erklärt wird. Es folgt, dass notwendig ist, dass niemand vollständig erklärt, warum diese Wahrheit wahr ist und von niemandem vollständig erklärt wird:

$$\Box\sim\!\mathcal{E}(p \,\&\, \sim\!\mathcal{E} p) \qquad\qquad \text{(Gl. 9.24–9.26, RN)} \qquad\qquad (9.27)$$

Es ist dann aber unmöglich, dass jemand vollständig erklärt, warum diese Wahrheit wahr ist und von niemandem vollständig erklärt wird:

$$\sim\!\Diamond\mathcal{E}(p \ \& \sim\!\mathcal{E}p) \qquad\qquad \text{(Gl. 9.27, DEF-}\Diamond) \qquad (9.28)$$

Es ist also widersprüchlich anzunehmen, dass diese Wahrheit eine Wahrheit über den Bereich β ist und von niemandem vollständig erklärt wird. Denn wenn diese Wahrheit eine Wahrheit über den Bereich β ist und von niemandem vollständig erklärt wird, dann ist es möglich und unmöglich, dass jemand vollständig erklärt, warum diese Wahrheit wahr ist und von niemandem vollständig erklärt wird:

$$\Diamond\mathcal{E}(p \ \& \sim\!\mathcal{E}p) \ \& \sim\!\Diamond\mathcal{E}(p \ \& \sim\!\mathcal{E}p) \qquad\qquad \text{(Gl. 9.23 und 9.28)} \qquad (9.29)$$

Es folgt, dass es falsch ist, dass diese Wahrheit eine Wahrheit über den Bereich β ist und von niemandem vollständig erklärt wird, wenn alle Wahrheiten über den Bereich β von jemandem vollständig erklärt werden können:

$$\forall p(\beta p \rightarrow \Diamond\mathcal{E}p) \rightarrow \sim\!(\beta p \ \& \sim\!\mathcal{E}p) \qquad\qquad \text{(Gl. 9.19-9.29)} \qquad (9.30)$$

Es folgt, dass nicht alle Wahrheiten über den Bereich β von jemandem vollständig erklärt werden können, wenn es (wenigstens) eine Wahrheit über den Bereich β gibt, die nicht von jemandem vollständig erklärt wird:

$$\forall p(\beta p \rightarrow \Diamond\mathcal{E}p) \rightarrow \forall p\!\sim\!(\beta p \ \& \sim\!\mathcal{E}p) \qquad\qquad \text{(Gl. 9.19–9.30, INTRO-}\forall) \qquad (9.31)$$

$$\sim\!\forall p\!\sim\!(\beta p \ \& \sim\!\mathcal{E}p) \rightarrow \sim\!\forall p(\beta p \rightarrow \Diamond\mathcal{E}p) \qquad\qquad \text{(Gl. 9.31)} \qquad (9.32)$$

$$\exists p(\beta p \ \& \sim\!\mathcal{E}p) \rightarrow \sim\!\forall p(\beta p \rightarrow \Diamond\mathcal{E}p) \qquad\qquad \text{(Gl. 9.32, DEF-}\exists) \qquad (9.33)$$

Nun gibt es aber (wenigstens) eine Wahrheit über den Bereich β, die nicht von jemandem vollständig erklärt wird. Es folgt, dass nicht alle Wahrheiten über den Bereich β von jemandem vollständig erklärt werden können:

$$\sim\!\forall p(\beta p \rightarrow \Diamond\mathcal{E}p) \qquad\qquad \text{(Gl. 9.18 und 9.33)} \qquad (9.34)$$

Es gibt also wenigstens eine Wahrheit über den Bereich β, die nicht von jemandem vollständig erklärt werden kann:

$$\exists p(\beta p \ \& \sim\!\Diamond\mathcal{E}p) \qquad\qquad \text{(Gl. 9.34, DEF-}\exists) \qquad (9.35)$$

Was für einen beliebigen Bereich gilt, gilt für jeden Bereich: Es gibt in jedem Bereich wenigstens eine unerklärliche Wahrheit. QED.

9.6.3   Es gibt völlig unerklärliche Wahrheiten

Um in einer formallogischen Sprache zu zeigen, dass es völlig unerklärliche Wahrheiten gibt, bedarf es lediglich einer zusätzlichen Abkürzung und der Einführung eines zusätzlichen erkenntnistheoretischen Prinzips: Wenn $\ulcorner \varepsilon A \urcorner$ eine Abkürzung für $\ulcorner$ es gibt jemanden, der zumindest zum Teil erklärt, warum A $\urcorner$ ist, dann genügt es, anzunehmen, dass gilt: Wenn jemand zumindest zum Teil erklärt, warum A, dann A:

$$(\text{FAKT}-\varepsilon) \qquad\qquad \vdash \varepsilon A \to A$$

Denn dann muss man lediglich voraussetzen, dass es (wenigstens) eine Wahrheit gibt, die nicht im Rahmen einer Erklärung erwähnt wird (das heißt eine Wahrheit, die nicht Konjunkt einer zumindest zum Teil erklärten Konjunktion ist):

$$\exists p(p \ \& \ {\sim}\exists q \varepsilon(p \ \& \ q)) \qquad\qquad \text{(Prämisse)} \qquad (9.36)$$

Um dann zu zeigen, dass es völlig unerklärliche Wahrheiten gibt, nehmen wir an, dass alle Wahrheiten von jemandem zumindest zum Teil erklärt werden können, und folgern, dass von jemandem zumindest zum Teil erklärt werden kann, warum diese Wahrheit wahr ist und nicht im Rahmen einer Erklärung erwähnt wird:

$$\forall p(p \to \Diamond \varepsilon p) \qquad\qquad\qquad\qquad\quad \text{(Annahme)} \qquad (9.37)$$

$$p \ \& \ {\sim}\exists q \varepsilon(p \ \& \ q) \qquad\qquad\qquad\quad \text{(Annahme)} \qquad (9.38)$$

$$p \ \& \ {\sim}\exists q \varepsilon(p \ \& \ q) \to \Diamond \varepsilon(p \ \& \ {\sim}\exists q \varepsilon(p \ \& \ q)) \qquad \text{(Gl. 9.37, ELIM-}\forall) \qquad (9.39)$$

$$\Diamond \varepsilon(p \ \& \ {\sim}\exists q \varepsilon(p \ \& \ q)) \qquad\qquad\qquad \text{(Gl. 9.38 und 9.39)} \qquad (9.40)$$

Es ist also möglich, dass jemand zumindest zum Teil erklärt, warum diese Wahrheit wahr ist und nicht im Rahmen einer Erklärung erwähnt wird. Nun nehmen wir aber einmal an, jemand würde zumindest zum Teil erklären, warum diese Wahrheit wahr ist und nicht im Rahmen einer Erklärung erwähnt wird:

$$\varepsilon(p \ \& \ {\sim}\exists q \varepsilon(p \ \& \ q)) \qquad\qquad \text{(Annahme)} \qquad (9.41)$$

Es wäre dann wahr, dass diese Wahrheit im Rahmen einer Erklärung erwähnt wird:

$$\forall q {\sim} \varepsilon(p \ \& \ q) \to {\sim} \varepsilon(p \ \& \ {\sim}\exists q \varepsilon(p \ \& \ q)) \qquad\quad \text{(ELIM-}\forall) \qquad (9.42)$$

$${\sim}{\sim} \varepsilon(p \ \& \ {\sim}\exists q \varepsilon(p \ \& \ q)) \to {\sim}\forall q {\sim} \varepsilon(p \ \& \ q) \qquad \text{(Gl. 9.42)} \qquad (9.43)$$

$${\sim}{\sim} \varepsilon(p \ \& \ {\sim}\exists q \varepsilon(p \ \& \ q)) \qquad\qquad\qquad \text{(Gl. 9.41)} \qquad (9.44)$$

$${\sim}\forall q {\sim} \varepsilon(p \ \& \ q) \qquad\qquad\qquad \text{(Gl. 9.43 und 9.44)} \qquad (9.45)$$

$$\exists q \varepsilon(p \ \& \ q) \qquad\qquad\qquad \text{(Gl. 9.45, DEF-}\exists) \qquad (9.46)$$

Es wäre dann aber auch nicht wahr, dass diese Wahrheit im Rahmen einer Erklärung erwähnt wird:

$$\sim\exists q\varepsilon(p \mathbin\& q) \qquad\qquad \text{(Gl. 9.41, FAKT-}\varepsilon\text{)} \qquad\qquad (9.47)$$

Es wäre dann also wahr und nicht wahr, dass diese Wahrheit im Rahmen einer Erklärung erwähnt wird:

$$\exists q\varepsilon(p \mathbin\& q) \mathbin\& \sim\exists q\varepsilon(p \mathbin\& q) \qquad\qquad \text{(Gl. 9.46 und 9.47)} \qquad\qquad (9.48)$$

Es ist also widersprüchlich anzunehmen, jemand würde zumindest zum Teil erklären, warum diese Wahrheit wahr ist und nicht im Rahmen einer Erklärung erwähnt wird. Es folgt, dass notwendig ist, dass niemand zumindest zum Teil erklärt, warum diese Wahrheit wahr ist und nicht im Rahmen einer Erklärung erwähnt wird:

$$\Box\sim\varepsilon(p \mathbin\& \sim\exists q\varepsilon(p \mathbin\& q)) \qquad\qquad \text{(Gl. 9.41–9.48, RN)} \qquad\qquad (9.49)$$

Es ist dann aber unmöglich, dass jemand zumindest zum Teil erklärt, warum diese Wahrheit wahr ist und nicht im Rahmen einer Erklärung erwähnt wird:

$$\sim\Diamond\varepsilon(p \mathbin\& \sim\exists q\varepsilon(p \mathbin\& q)) \qquad\qquad \text{(Gl. 9.49, DEF-}\Diamond\text{)} \qquad\qquad (9.50)$$

Es ist also widersprüchlich anzunehmen, dass diese Wahrheit wahr ist und nicht im Rahmen einer Erklärung erwähnt wird. Denn wenn diese Wahrheit wahr ist und nicht im Rahmen einer Erklärung erwähnt wird, dann ist es möglich und unmöglich, dass jemand zumindest zum Teil erklärt, warum diese Wahrheit wahr ist und nicht im Rahmen einer Erklärung erwähnt wird:

$$\Diamond\varepsilon(p \mathbin\& \sim\exists q\varepsilon(p \mathbin\& q)) \mathbin\& \sim\Diamond\varepsilon(p \mathbin\& \sim\exists q\varepsilon(p \mathbin\& q)) \quad \text{(Gl. 9.40 und 9.50)} \quad (9.51)$$

Es folgt, dass es falsch ist, dass diese Wahrheit wahr ist und nicht im Rahmen einer Erklärung erwähnt wird, wenn alle Wahrheiten von jemandem zumindest zum Teil erklärt werden können:

$$\forall p(p \rightarrow \Diamond\varepsilon p) \rightarrow \sim(p \mathbin\& \sim\exists q\varepsilon(p \mathbin\& q)) \qquad\qquad \text{(Gl. 9.37–9.51)} \qquad\qquad (9.52)$$

Es folgt, dass nicht alle Wahrheiten von jemandem zumindest zum Teil erklärt werden können, wenn es (wenigstens) eine Wahrheit gibt, die nicht im Rahmen einer Erklärung erwähnt wird:

$$\forall p(p \rightarrow \Diamond\varepsilon p) \rightarrow \forall p\sim(p \mathbin\& \sim\exists q\varepsilon(p \mathbin\& q)) \qquad \text{(Gl. 9.37–9.52, INTRO-}\forall\text{)} \qquad (9.53)$$

$$\sim\forall p\sim(p \mathbin\& \sim\exists q\varepsilon(p \mathbin\& q)) \rightarrow \sim\forall p(p \rightarrow \Diamond\varepsilon p) \qquad \text{(Gl. 9.53)} \qquad (9.54)$$

$$\exists p(p \mathbin\& \sim\exists q\varepsilon(p \mathbin\& q)) \rightarrow \sim\forall p(p \rightarrow \Diamond\varepsilon p) \qquad \text{(Gl. 9.54, DEF-}\exists\text{)} \qquad (9.55)$$

Nun gibt es aber (wenigstens) eine Wahrheit, die nicht im Rahmen einer Erklärung erwähnt wird. Es folgt, dass nicht alle Wahrheiten von jemandem zumindest zum Teil erklärt werden können:

$$\sim\forall p(p \to \Diamond \varepsilon p) \qquad \text{(Gl. 9.36 und 9.55)} \qquad (9.56)$$

Es gibt also wenigstens eine Wahrheit, die nicht von jemandem zumindest zum Teil erklärt werden kann:

$$\exists p(p \ \& \sim\Diamond \varepsilon p) \qquad \text{(Gl. 9.56, DEF-}\exists) \qquad (9.57)$$

Es gibt also wenigstens eine völlig unerklärliche Wahrheit. QED.

9.6.4 Wer alle unerklärlichen Wahrheiten kennt, kennt alle Wahrheiten

Um in einer formallogischen Sprache zu zeigen, dass man alle Wahrheiten kennt, wenn man alle unerklärlichen Wahrheiten kennt, bedarf es lediglich einer zusätzlichen Abkürzung, der Einführung eines zusätzlichen logischen Prinzips und der Einführung eines zusätzlichen erkenntnistheoretischen Prinzips: Wenn $\ulcorner Kp \urcorner$ eine Abkürzung für $\ulcorner$ich weiß, dass $p \urcorner$ ist, dann bedarf es lediglich der Annahme, dass das folgende, in der Modallogik übliche, logische Prinzip gültig ist (Vgl. z. B. Kripke 1959, S. 1):

$$\text{(K)} \qquad \vdash \Box(p \to q) \to (\Box p \to \Box q) \qquad (9.58)$$

Es bedarf außerdem der Annahme, dass das folgende, in der epistemischen Logik übliche, erkenntnistheoretische Prinzip gültig ist[17]:

$$\text{(KON-}K) \qquad \vdash K(p \ \& \ q) \to (Kp \ \& \ Kq) \qquad (9.59)$$

Wie wir bereits gesehen haben, gibt es (wenigstens) eine unerklärliche Wahrheit:

$$\exists p(p \ \& \sim\Diamond \mathcal{E} p) \qquad \text{(Prämisse)} \qquad (9.60)$$

Nun nehmen wir aber an, ich würde alle unerklärlichen Wahrheiten kennen:

$$\forall p((p \ \& \sim\Diamond \mathcal{E} p) \to Kp) \qquad \text{(Annahme)} \qquad (9.61)$$

---

[17]Für eine ausführliche Verteidigung dieses erkenntnistheoretischen Prinzips vgl. Williamson (2000, S. 275–282).

Die Konjunktion einer unerklärlichen Wahrheit mit einer beliebigen Wahrheit ist selbst eine unerklärliche Wahrheit:

$$p \,\&\, {\sim}\Diamond\mathcal{E}p \qquad \text{(Annahme)} \qquad (9.62)$$

$$q \qquad \text{(Annahme)} \qquad (9.63)$$

$$\mathcal{E}(p \,\&\, q) \to \mathcal{E}p \qquad \text{(KON-}\mathcal{E}) \qquad (9.64)$$

$$\sim\mathcal{E}p \to {\sim}\mathcal{E}(p \,\&\, q) \qquad \text{(Gl.\,9.64)} \qquad (9.65)$$

$$\Box({\sim}\mathcal{E}p \to {\sim}\mathcal{E}(p \,\&\, q)) \qquad \text{(Gl.\,9.64 und 9.65, RN)} \qquad (9.66)$$

$$\Box{\sim}\mathcal{E}p \qquad \text{(Gl.\,9.62, DEF-}\Diamond) \qquad (9.67)$$

$$\Box{\sim}\mathcal{E}(p \,\&\, q) \qquad \text{(Gl.\,9.66 und 9.67, K)} \qquad (9.68)$$

$$\sim\Diamond\mathcal{E}(p \,\&\, q) \qquad \text{(Gl.\,9.68, DEF-}\Diamond) \qquad (9.69)$$

$$(p \,\&\, q) \,\&\, {\sim}\Diamond\mathcal{E}(p \,\&\, q) \qquad \text{(Gl.\,9.62, 9.63 und 9.69)} \qquad (9.70)$$

Wenn die Konjunktion dieser unerklärlichen Wahrheit mit der beliebigen Wahrheit aber selbst eine unerklärliche Wahrheit ist, dann kenne ich die Konjunktion dieser unerklärlichen Wahrheit mit der beliebigen Wahrheit:

$$((p \,\&\, q) \,\&\, {\sim}\Diamond\mathcal{E}(p \,\&\, q)) \to K(p \,\&\, q) \qquad \text{(Gl.\,9.61 ELIM-}\forall) \qquad (9.71)$$

$$K(p \,\&\, q) \qquad \text{(Gl.\,9.70 und 9.71)} \qquad (9.72)$$

Wenn ich die Konjunktion dieser unerklärlichen Wahrheit mit der beliebigen Wahrheit kenne, dann kenne ich aber auch die beliebige Wahrheit:

$$K q \qquad \text{(Gl.\,9.72, KON-}K) \qquad (9.73)$$

Ich kenne also jede beliebige Wahrheit:

$$q \to K q \qquad \text{(Gl.\,9.63–9.73)} \qquad (9.74)$$

Es folgt, dass, wenn es eine Wahrheit gibt, die von niemandem vollständig erklärt werden kann, dass ich dann alle Wahrheiten kenne, wenn ich alle unerklärlichen Wahrheiten kenne:

$$(p \,\&\, {\sim}\Diamond\mathcal{E}p) \to (q \to K q)$$
$$\text{(Gl.\,9.62–9.74)} \qquad (9.75)$$

$$\sim(q \to K q) \to {\sim}(p \,\&\, {\sim}\Diamond\mathcal{E}p)$$
$$\text{(Gl.\,9.75)} \qquad (9.76)$$

$$\forall p((p \,\&\, {\sim}\Diamond\mathcal{E}p) \to K p) \to ({\sim}(q \to K q) \to {\sim}(p \,\&\, {\sim}\Diamond\mathcal{E}p))$$
$$\text{(Gl.\,9.61–9.76)} \qquad (9.77)$$

$$(\forall p((p \,\&\, {\sim}\Diamond\mathcal{E}p) \to K p) \,\&\, {\sim}(q \to K q)) \to {\sim}(p \,\&\, {\sim}\Diamond\mathcal{E}p)$$
$$\text{(Gl.\,9.77)} \qquad (9.78)$$

$$(\forall p((p \,\&\, \sim\lozenge\mathcal{E}p) \to Kp) \,\&\, \sim(q \to Kq)) \to \forall p\sim(p \,\&\, \sim\lozenge\mathcal{E}p)$$
(Gl. 9.61–9.78, INTRO-$\forall$) (9.79)

$$\sim\forall p\sim(p \,\&\, \sim\lozenge\mathcal{E}p) \to \sim(\forall p((p \,\&\, \sim\lozenge\mathcal{E}p) \to Kp) \,\&\, \sim(q \to Kq))$$
(Gl. 9.79) (9.80)

$$\exists p(p \,\&\, \sim\lozenge\mathcal{E}p) \to (\forall p((p \,\&\, \sim\lozenge\mathcal{E}p) \to Kp) \to (q \to Kq))$$
(Gl. 9.80, DEF-$\exists$) (9.81)

$$(\exists p(p \,\&\, \sim\lozenge\mathcal{E}p) \,\&\, \forall p((p \,\&\, \sim\lozenge\mathcal{E}p) \to Kp)) \to (q \to Kq)$$
(Gl. 9.81) (9.82)

$$(\exists p(p \,\&\, \sim\lozenge\mathcal{E}p) \,\&\, \forall p((p \,\&\, \sim\lozenge\mathcal{E}p) \to Kp)) \to \forall q(q \to Kq)$$
(Gl. 9.61–9.82, INTRO-$\forall$) (9.83)

$$\exists p(p \,\&\, \sim\lozenge\mathcal{E}p) \to (\forall p((p \,\&\, \sim\lozenge\mathcal{E}p) \to Kp) \to \forall q(q \to Kq))$$
(Gl. 9.83) (9.84)

Nun gibt es aber eine Wahrheit, die von niemandem vollständig erklärt werden kann. Es folgt, dass ich alle Wahrheiten kenne, wenn ich alle unerklärlichen Wahrheiten kenne:

$$\forall p((p \,\&\, \sim\lozenge\mathcal{E}p) \to Kp) \to \forall q(q \to Kq) \qquad \text{(Gl. 9.60 und 9.84)} \quad (9.85)$$

Was für mich gilt, gilt für alle: Wer alle unerklärlichen Wahrheiten kennt, kennt alle Wahrheiten. QED.

Literatur

Boolos, G. (1996). *The logic of provability*. Cambridge: Cambridge University Press.

Cantor, G. (1892). Über eine elementare Frage der Mannigfaltigkeitslehre. *Jahresbericht der Deutschen Mathematiker Vereinigung, 1,* 75–78.

Fine, K. (2012). A guide to ground. In von F. Correia & B. Schnieder (Hrsg.), *Metaphysical grounding. understanding the structure of reality* (S. 37–80). Cambridge: Cambridge University Press.

Fitch, F. B. (1963). A logical analysis of some value concepts. *Journal of Symbolic Logic, 28,* 135–142.

Gödel, K. (1931). Über formal unentscheidbare Sätze der Principia Mathematica und verwandter Systeme I. *Monatshefte für Mathematik, 38,* 173–198.

Kripke, S. (1959). A completeness theorem in modal logic. *Journal of Symbolic Logic, 24,* 1–14.

Salerno, J. (2009). Knowability Noir: 1945–1965. In von J. Salerno (Hrsg.), *New essays on the knowability paradox* (S. 29–48). Oxford: Oxford University Press.

Salerno, J. (2018). Knowability and a new paradox of happiness. In von H. Van Ditmarsch & G. Sandu (Hrsg.), *Jaakko Hintikka on knowledge and game-theoretical semantics* (S. 457–474). Berlin: Springer.

Schnieder, B. (2011). A logic for 'because'. *The Review of Symbolic Logic, 4,* 445–465.

Williamson, T. (2000). *Knowledge and its limits*. Oxford: Oxford University Press.

Williamson, T. (2007). *The philosophy of philosophy*. Oxford: Blackwell.

Gödel, mathematischer Realismus und Antireduktionismus

10

Reinhard Kahle

Es ist bekannt, dass Kurt Gödel (2003, S. 447) im „mathematischen Realismus" seine philosophische Grundposition wiedergegeben sah.

Zielsetzung dieses Essays ist es zu zeigen, dass sich auch umgekehrt ein mathematischer Realismus praktisch zwangsläufig als eine Konsequenz aus dem (ersten) Unvollständigkeitssatz ergibt. Im Weiteren zeigt sich damit eine unüberwindliche Hürde für eine reduktionistische Mathematik, die z. B. die reellen Zahlen auf die natürlichen Zahlen zurückführen will.

10.1 Historischer Kontext

Ausgangspunkt für die mathematischen Forschungen Gödels war das *Hilbertsche Programm.* Als Antwort auf die in der naiven Mengenlehre auftretenden Paradoxien einerseits und auf die versuchten Eingriffe der Philosophie in die Mathematik durch (anfänglich) Emil du Bois-Reymond und (später) Bertus Brouwer andererseits, hatte

Diese Arbeit wurde u. a. von der Udo Keller-Stiftung und durch die Portugiesische Forschungsgemeinschaft FCT über das *Centro de Matemática e Aplicações,* UID/MAT/00297/2020, gefördert.

R. Kahle (✉)
Theorie und Geschichte der Wissenschaften, Universität Tübingen, Tübingen, Deutschland
E-mail: kahle@mat.uc.pt

CMA, FCT, Universidade Nova de Lisboa, Caparica, Portugal

O. Passon und C. Benzmüller (Hrsg.), *Wider den Reduktionismus,*
https://doi.org/10.1007/978-3-662-63187-4_10

David Hilbert ein reduktionistisches Programm entworfen, in dessen Rahmen höhere Mathematikkonzepte auf *finite Mathematik* zurückgeführt werden sollten.[1]

Als finite Mathematik wurde dabei ein unkontroverser Teil der Mathematik aufgefasst, der sich, vereinfacht gesagt, auf die natürliche Zahlen und quantorenfreie Aussagen über diese beschränkte. Insbesondere sollte diese finite Mathematik für die Intuitionisten um Brouwer akzeptabel sein. Der Hilbertsche Reduktionismus war in dieser Hinsicht rein taktisch:[2] Sollte die angestrebte Reduktion gelingen, würden die höheren Mathematikkonzepte eine finite Rechtfertigung erfahren, und der Intuitionist sollte verpflichtet sein, diese anzuerkennen.

10.2   Gödels Unvollständigkeitssätze

Der zweite Gödelsche Unvollständigkeitssatz zeigt in aller Deutlichkeit, dass sich das Hilbertsche Programm in seiner ursprünglichen Intention nicht durchführen lässt. Für unser Anliegen hier ist allerdings der erste Unvollständigkeitssatz interessanter. Dafür sei T eine formale Theorie, die ausreichend Arithmetik enthält und gewisse technische Voraussetzungen erfüllt, u. a. dass sie konsistent ist. Dann besagt der erste Gödelsche Unvollständigkeitssatz bekanntlich, dass T syntaktisch unvollständig ist.[3] Das heißt es gibt einen Formel ϕ_G der zugrunde liegenden Sprache, sodass

$$T \nvdash \phi_G \quad \text{und} \quad T \nvdash \neg\phi_G.$$

Die Stärke von Gödels Resultat liegt darin, dass das Resultat generisch ist: Auch für stärkere Theorien, in denen die in T unabhängige Formel ϕ_G beweisbar ist (z. B. $T' := T + \{\phi_G\}$), lässt sich immer eine neue (natürlich von ϕ_G verschiedene) Formel ϕ_G' finden, die sich als von der neuen Theorie unabhängig erweisen lässt.

10.3   Das realistische Korollar

Wir nehmen im Folgenden T und ϕ_G wie oben an. Aus der Voraussetzung, dass T konsistent ist, und dem Umstand, dass $T \nvdash \neg\phi_G$, folgt unmittelbar, dass $T + \{\phi_G\}$ konsistent ist. In dem Maße, in dem man geneigt ist, T als Formalisierung eines mathematischen Gegenstandsbereichs zu betrachten, ergibt sich, dass auch $T' :=$

---

[1]Zu den Paradoxien siehe z. B. Hilbert (1905); eine ausführliche Diskussion findet sich in Kahle (2006). Die Kritik an du Bois-Reymond wurde mit dem Ausspruch „In der Mathematik gibt es kein Ignorabimus!" in der berühmten Rede auf dem Internationalen Mathematikerkongress in Paris deutlich zum Ausdruck gebracht (Hilbert 1900); die Kritik an Brouwers Intuitionismus durchzieht die ganze Entwicklung der Beweistheorie (z. B. Hilbert 1922).

[2]Smorynski (2002) betrachtet Hilberts Position als „strategisch".

[3]Neben der Originalarbeit (Gödel 1931) kann man für eine moderne Darstellung z. B. Smorynski (1977) heranziehen; eine informelle Beschreibung der Gödelschen Sätze wurde auch in Kahle (2007) gegeben.

$T + \{\phi_G\}$ aufgrund seiner Konsistenz einen mathematischen Gegenstandsbereich formalisiert. Und das Gleiche gilt aus analogem Grund auch für $T'' := T + \{\neg\phi_G\}$. Der durch ϕ_G ausgedrückte mathematische Sachverhalt, sei er nun wahr oder falsch im intendierten Modell, wird also von T nicht erfasst. Da ϕ_G eine Existenzaussage sein kann, können wir auch sagen, dass die Gegenstandsbereiche von T, T' und T'' insofern verschieden sind, als T' und T'' sich um den durch ϕ_G charakterisierten „Gegenstand" unterscheiden und T offenlässt, ob dieser miterfasst ist oder nicht. In der Regel wird man einem der beiden Gegenstandsbereiche aus mathematischen Gründen den Vorzug geben. Es kann aber auch vorkommen, wie im Fall der Kontinuumshypothese, dass man nicht genügend Evidenz hat, sich eindeutig entscheiden zu können. Doch unabhängig von der Wahl, die man an dieser Stelle zu treffen hat, ist klar, dass sowohl T' als auch T'' ihre – durchaus wörtlich gemeinte – Existenzberechtigung haben. Das bedeutet aber, dass es für jede mathematische Theorie, auf die der erste Gödelsche Unvollständigkeitssatz anwendbar ist, eine Erweiterung gibt, die sich auf einen erweiterten mathematischen Gegenstandsbereich bezieht.[4]

In diesem Sinne zeigt der erste Gödelsche Unvollständigkeitssatz, dass es immer wieder „neue" mathematische Entitäten gibt, die sich durch eine gegebene formale Theorie nicht erfassen lassen.[5]

10.4   Das antireduktionistische Korollar

Ein mathematischer Reduktionismus beinhaltet, dass sich gewisse (in der Regel: „höhere") mathematische Konzepte auf andere („niedere") zurückführen lassen. Eine solche Reduktion ist in Einzelfällen durchaus möglich, z. B. im Fall der Kodierung

---

[4]Für die Theorie, in der die der mathematischen Intuition widerstrebende Formel ϕ_G oder $\neg\phi_G$ gilt, erhalten wir sogenannte „Nichtstandardmodelle"; diese sind für die Peano-Arithmetik (PA) gut erforscht und verstanden. Sie waren schon Dedekind bekannt (siehe Tapp 2017 und Kahle 2017) und Gödel selbst hat sie nach Takeuti wie folgt charakterisiert:

[Gödel's] way of teaching nonstandard models was an interesting one. It went as follows. Let T be a theory with a nonstandard model. By virtue of his Incompleteness Theorem, the consistency proof of T cannot be carried out within T. Consequently, T and the proposition "T is inconsistent" is consistent. There is, therefore, a natural number N which is the Gödel number of the proof leading to a contradiction from T. Such a number is obviously an infinite natural number. (Yasugi und Passell 2003, S. 3)

Für derartige Modelle der Arithmetik haben Kikuchi und Kurahashi (2016) die passende Bezeichnung „*insane* models of PA" eingeführt.

[5]„Formale Theorie" ist hier, wie heute allgemein üblich, auf eine Theorie in der *Logik erster Stufe* bezogen. Während man durchaus argumentieren kann, dass die Sachlage für mathematische Entitäten in einer Logik höherer Stufe anders liegen könnte, würde dies aber um den – sehr viel höheren – Preis erkauft, dass sich die *Logik zweiter Stufe* bereits nicht mehr axiomatisieren lässt. Auch das ist ein Korollar aus dem (Beweis des) ersten Gödelschen Unvollständigkeitssatz(es); siehe dazu auch Kahle (2019).

der komplexen Zahlen durch Paare reeller Zahlen, mit entsprechend eingeführten Operationen auf diesen Zahlenpaaren für die komplexe Addition und Multiplikation. Die analytische Geometrie, die es erlaubt, die Aussagen der euklidischen Geometrie in analytische Ausdrücke im $\mathbb{R}^2$ bzw. $\mathbb{R}^3$ zu übersetzen, ist vielleicht das beeindruckenste reduktionistische Resultat in der Mathematik. Dennoch ergibt sich nach Gödel, dass einem universellen reduktionistischen Programm in der Mathematik prinzipielle Grenzen gesetzt sind. Dies zeigt sich in erster Linie bei dem Versuch einer Rückführung der reellen Zahlen auf die natürlichen Zahlen.

Für eine derartige Rückführung müsste man in der Lage sein, Aussagen über reelle Zahlen immer in (eventuell sehr viel kompliziertere) Aussagen über natürliche Zahlen zu übersetzen. Nun ergibt sich schon für die oben betrachtete Formel ϕ_G, dass der durch sie ausgedrückte mathematische Sachverhalt nicht in der ursprünglichen Theorie T entschieden werden kann. Heute wissen wir aus eingehenden formalen Studien über arithmetische Theorien und Theorien für Subsysteme der Analysis, dass es sich bei den zur Diskussion stehenden Sachverhalten keinesfalls nur um „pathologische Grenzfälle" handelt, sondern um Sachverhalte mit klarem mathematischen Gehalt.[6] Aus den dabei gewonnenen Separationsresultaten, die in ihrem Kern immer auf den ersten Gödelschen Unvollständigkeitssatz zurückverweisen, ergibt sich insbesondere, dass sich die Analysis nicht auf die Arithmetik reduzieren lässt.

Allgemein lässt sich sagen, dass es zu jeder hinreichend starken Theorie T, auf die ein Reduktionist höhere Mathematik reduzieren will, eine Formel ϕ_G gibt, die einen Sachverhalt dieser höheren Mathematik ausdrückt, der durch T nicht erfasst wird. An der Frage, wie der Reduktionist diesen Aspekt des ersten Gödelschen Unvollständigkeitssatz umgehen könnte, wird er aus formalen Gründen scheitern.

Literatur

Gödel, K. (1931). Über formal unentscheidbare Sätze der Principia Mathematica und verwandter Systeme. *Monatshefte für Mathematik und Physik, 38,* 173–198.

Gödel, K. (2003). Grandjean's questionnaire. In von S. Feferman (Hrsg.), *Collected work* (Bd. IV, S. 446–449). Oxford University Press: Oxford.

Hilbert, D. (1900). Mathematische Probleme. Vortrag, gehalten auf dem internationalen Mathematiker-Kongreß zu Paris 1900. In *Nachrichten von der königl. Gesellschaft derWissenschaften zu Göttingen. Mathematisch-physikalische Klasse aus dem Jahre 1900* (S. 253–297).

Hilbert, D. (1905). Über die Grundlagen der Logik und der Arithmetik. In von A. Krazer (Hrsg.), *Verhandlungen des Dritten Internationalen Mathematiker-Kongresses in Heidelberg vom 8. Bis 13. August 1904* (S. 174–185). Leipzig.

Hilbert, D. (1922). Neubegründung der Mathematik. *Abhandlungen aus dem Mathematischen Seminar der Hamburgischen Universität, 1,* 157–177.

---

[6]Wir können hier einerseits auf das Programm der *reversen Mathematik* (Simpson 2009) verweisen, das konkrete mathematische Theoreme im Hinblick auf ihre beweistheoretische Stärke hin untersucht; andererseits erlaubt die durch Arbeiten von Gentzen motivierte *Ordinalzahlanalyse* (Rathjen 2006), ein konkretes (ordinales) Maß für die beweistheoretische Stärke von Theorien zu bestimmen.

Kahle, R. (2006). *David Hilbert über Paradoxien*. Techn. Ber. 06–17. Departamento de Matemática, Universidade de Coimbra.

Kahle, R. (2007). Die Gödelschen Unvollständigkeitssätze. *Mathematische Semesterberichte, 54*(1), (1–12). (Nachgedruckt in Facetten der Mathematik, Springer 2018).

Kahle, R. (2017). Von Dedekind zu Zermelo versus Peano zu Gödel. *Mathematische Semesterberichte, 64*(2), (159–167). (Auch veröffentlicht in dem Band In *Memoriam Richard Dedekind (1831–1916)*, WTM, Verlag für wissenschaftliche Texte und Medien, Münster, 2017).

Kahle, R. (2019). Is there a "Hilbert thesis"? *Studia Logica, 107*(1), (145–165) (Special Issue on general proof theory. T. Piecha & P. Schroeder-Heister (Guest Editors)).

Kikuchi, M., & Kurahashi, T. (2016). Illusory models of Peano arithmetic. *The Journal of Symbolic Logic, 81*(3), 1163–1175.

Rathjen, M. (2006). The art of ordinal analysis. In Proceedings of the International Congress of Mathematicians (ICM), Madrid, Spain, August 22–30. *Bd. II: Invited lectures* (S. 45–69). Zürich: European Mathematical Society (EMS).

Simpson, S. G. (2009). *Subsystems of second order arithmetic* (2 Aufl.). Cambridge University Press: Cambridge.

Smorynski, C. (1977). The incompleteness theorems. In von J. Barwise (Hrsg.), *Handbook of mathematical logic* (S. 821–865). Amsterdam: North-Holland.

Smorynski, C. (2002). Gödels Unvollständigkeitssätze. In von B. Buldt, E. Köhler, M. Stöltzner, P. Weibel, C. Klein, & W. DePauli-Schimanovich-Göttig (Hrsg.), *Kurt Gödel, Wahrheit und Beweisbarkeit: Bd. II. Kompendium zum Werk. öbv & hpt* (S. 147–159).

Tapp, C. (2017). Richard Dedekind: Brief an keferstein. In von K. Scheel, T. Sonar, & P. Ullrich (Hrsg.), *In memoriam Richard Dedekind (1831–1916)* (S. 205–211). Münster: WTM, Verlag für wissenschaftliche Texte und Medien.

Yasugi, M., & Passell, N. (Hrsg.). (2003). *Memoirs of a proof theorist* (English translation of a collection of essays written by Gaisi) Takeuti: World Scientific.